# Medical Physics

**SECOND EDITION**

**University of Bath • Science 16–19**

Series Director: Prof. J. J. Thompson, CBE

# Medical Physics

**SECOND EDITION**

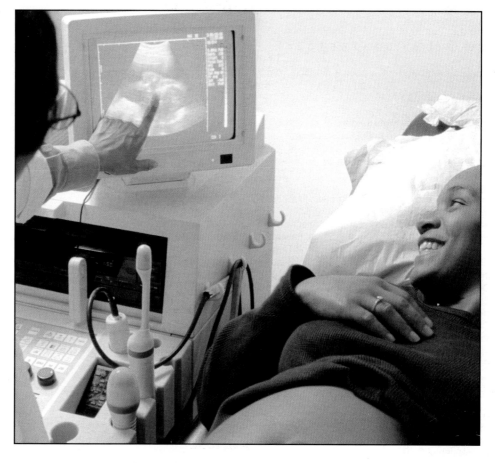

# MARTIN HOLLINS

First published in 1992 by:
Thomas Nelson and Sons Ltd

Second edition published in 2001 by:
Nelson Thornes Ltd
Delta Place
27 Bath Road
CHELTENHAM
GL53 7TH
United Kingdom

01 02 03 04 05 / 10 9 8 7 6 5 4 3 2 1

A catalogue record for this book is available from the British Library

ISBN 0 17 448253 1

New illustrations by Nigel Jordan
Page make-up by Jordan Publishing Design

Printed and bound in Italy by Canale

**Acknowledgements**
The authors and publishers are grateful to the following for permission to reproduce photographs:
Allsport **1.12**; Associated Press **10.2**; Atlantic Syndication Partners **4.9**; Brompton Hospital Cardiac Department **8.5**(**a**), (**b**); Eastman Dental Hospital **9.1**; Finetech Medical Ltd **5.14**; Getty Stone **p xii** (Lonnie Duke), **6.4** (Elie Bernager), **8.14** (Steven Pelets), **9.18** (RNHRD NHS Trust); Illustrated London News **p xiii top**; Impact Photos **p x** (Giles Barnard); Leeds University **2.18** (Dr P Evenett), Department of Pure and Applied Biology; Martyn Chillmaid **1.13**; Philip Harris Education **4.3b**, **6.1**; Pizza Express **2.13b**; Robert Harding Picture Library **p xi** (Richard C Novitz); Scala **2.12**(**a**), (**b**); Science Photolibrary **p xiii bottom**, **1.9** CNRI, **55** (James King-Holmes) **4.6**(**a**), (**b**), **5.9** (Simon Fraser), **7.5** Deep Light Productions, **7.8**, **10.5**, **10.7** CNRI, **11.5**, **12.12**, **12.13** BSIP Laurent/H.Americain **12.16** (Geoff Tompkinson), **13.1** (Geoff Tompkinson), **13.5** (Alfred Pasieka), **13.7** BSIP Ducloux, **13.8** (James Holmes) I Cell Tech; TRIP **8.13** Picturesque/Charles Gupton; University College London **9.21** (D. D'Sousa); Wellcome Trust Photolibrary **7.1**; William F Hinkes **9.19**, **12.17**; Wimpy **2.13**(**a**). Picture research by John Bailey.

AQA (NEAB) examination questions are reproduced by permission of the Assessment and Qualifications Alliance; Edexcel (London, Edexcel Specimen) are reproduced by permission of Edexcel and OCR (O & C, UCLES) are reproduced by permission of OCR.

Answers are the sole responsibility of the author and have not been provided or approved by the organisations. The organisations accept no responsibility whatsoever for the accuracy or method of working in the answers given.

Every effort has been made to trace all the copyright holders but if any have been overlooked the publishers will be pleased to make the necessary arrangements at the first opportunity.

# Contents

*Contents*

# Bath Advanced Science Series: Introduction

It is now some ten years since the **University of Bath 16–19 Project** was first published. The success of the material produced by the team of talented authors has been demonstrated consistently throughout that period. At the same time there has been a great deal of change in the provision of education, not only in the UK but throughout the world. With the ongoing drive towards the broadening of post-16 education – making it *more* accessible and *more* useful to *more* people – there is greater emphasis not only on expanding knowledge, but also on learning and understanding how this knowledge can be applied, and developing the skills required to do so.

The **Bath Advanced Science** series represents an extensive review, re-organisation, and update of the material produced in the original project. The author team has come together to draw on its collective expertise and experience to produce resources which reflect the demands of post-16 education today.

The series has been written to address the Advanced Subsidiary and Advanced GCE Science specifications introduced by the unitary awarding bodies from September 2000 following the Dearing *Review of Qualifications for 16–19 Year Olds* and the subsequent consultation *Qualifying for Success*.

■ The subject texts provide comprehensive coverage of the content required for the awarding body specifications in those subjects.
■ The topic books each provide detailed coverage which will support both option topics for AS and A GCE and other advanced level courses. They are also appropriate for students embarking on Foundation courses in Higher Education, or preparing for degrees in which these topics are studied.

The **Bath Advanced Science** series contains up-to-date material presented to encourage effective learning. The texts continue to be highly interactive – ensuring a degree of independent self-study and actively providing opportunities for students to acquire knowledge and understanding, develop skills and concepts, and to appreciate the applications and implications of science.

We have designed the **Bath Advanced Science** series specifically to help all students to develop the realisation that science *is* for everyone and to gain an appreciation and understanding of the importance and relevance of science to everyday life.

**Prof. J J Thompson**, CBE
Department of Education
University of Bath
*May 2000*

# How to use this book

## Introduction

**T**he aim of this book is to provide an introduction to medical physics which I hope you will find interesting and enjoyable. It is particularly geared to the requirements of students who are following an Advanced level or AS-level course in physics who are taking an option in medical physics.

The book covers the range of topics usually covered in medical physics though no syllabus or specification covers all of this. It is divided into themes which link to different parts of a basic study of physics. Each theme and most of the chapters are relatively self-contained, so that you can refer only to the chapters for the course you are following.

The links to basic physics are detailed in the prerequisites for a theme. The detailed content of each chapter is summarised in the learning objectives for each chapter. You should compare the chapter and section headings, together with the learning objectives, with your syllabus, to see which parts of the book are essential to your own course. The learning objectives are intended, also, to help you organise your learning and provide an aid to planning your study of the subject.

The book has been written so that there are a number of activities for you to carry out in addition to reading, which on its own is too passive to promote effective learning. In-text questions are designed to check your knowledge and understanding, as you proceed. Answers to these where appropriate, are given at the back of the book. Summary assignments cover larger topics, with the aim of helping you to consolidate and review the material. Some of the assignments will be examination questions which help you to see if your attainment is at the required level. Some questions do not have a simple, or even single, answer, and are intended for you to discuss with others.

Medical physics is not a fixed body of knowledge to be learnt, but a changing collection of theories and practices with vital implications for the health of human beings. A number of features of the book are designed to convey something of this variety and complexity. Extracts of articles on current developments are included to develop the skills of comprehension and evaluation of both technical and popular sources. Further reading suggestions, books, topical publications, such as journals and newspapers and websites can be found on www.bath-science.co.uk, the website for the series.

Investigations in the book are fairly open-ended; more like practical explorations of the ideas in topics, than standard laboratory procedures. Visits to appropriate hospital departments are invaluable for seeing actual diagnostic equipment in use, as this is rarely inexpensive enough for schools or colleges to purchase.

## Prerequisites

These are presented at the beginning of each theme, to remind you of the scientific content which underlies the material in the theme. You should usually have covered this in your previous GCSE or A-level studies. If you have not you will need to consult a suitable text, or your tutor.

## Learning objectives

These are statements of what you should have achieved when you have finished the chapter. They often link closely to statements in a course syllabus or specification and can be used to help you to make notes for revision, as well as for checking your progress.

## Questions

In-text questions may be answered from material just presented in the previous section, or may require additional thought and information.

## Summary assignments

These can be used to build up your general understanding of a topic and provide notes for reference when revising for examinations. Answering the questions and summary assignments successfully is the best way of knowing that you have achieved the required understanding – so don't miss them out!

## Case studies

Most of the time, we gain information not from text books but from a range of published material – can you comprehend a scientific paper, or evaluate the reliability of a newspaper article? These extracts can help.

## Investigations

What is the best way to measure our body temperature? How would we find out our exposure to radiation? What factors affect the clarity of an image? These are some of the questions which are used to further your skills and understanding in this subject.

## Further reading

More detailed information on medical physics topics and topical information can be found through the 'Bath' series website: www.bath-science.co.uk.

## Marginal notes

Marginal notes provide incidental background information along with examples and mini case studies which illustrate the topic.

# Introduction

**A** simple definition might be, physics applied to medicine, which then leads to the question 'what is medicine?' A description of medical practice would include the three stages of examination, diagnosis and treatment. Those matters directly concerned with the practice of medicine are often termed **clinical**, as distinct from the more general or theoretical **medical** aspects. For example a medical student usually undergoes a period of scientific pre-clinical education before getting involved with patients in clinical training. Physics is used in all three of these stages of medicine and in both clinical and more general roles.

**Examination of the human body** in health and disease is a basic part of the work of the physician and the physicist, which accounts for the similarity of the names, coming from the Greek *physike*, the science of nature.

The human body is composed of tissues with a wide range of physical functions, including mechanical, optical, acoustic, electrical, hydraulic, pneumatic and metabolic systems. There is scope therefore for measurement in almost any field of physics. This book reflects this; it includes a general review of the techniques of instrumentation in Chapter 4 and covers a wide range of examples from the clinical thermometer to the computed tomograph.

Hilary and Stephen Rose recently gave a series of lectures as Professors of **Physic**. They included topics of current concern around genetic engineering, such as designer babies and designer tomatoes. The Chelsea **physic** garden in London is full of medicinal plants, not Newtonian apple trees or Hawking black holes.

Medical examination of an athlete in training.

**Diagnosis** in medicine is based on the patient 'history' or physical examination, with an emphasis on what is abnormal or malfunctioning. The invention and development of new techniques for diagnosis, by physicists and engineers, working with clinicians, has to a large extent defined the practice of medical physics. It can be argued that its origin was the discovery of X-rays. Now the emphasis is on ultrasound, radiation protection, magnetic resonance, and extensive use of computers and microelectronics. Technology has enabled a better diagnosis to be made by offering a wider range of examination methods, leading to fuller information about the patient.

This is the part of medical physics most emphasised in this book, especially in Theme 3 which deals with some of the most advanced techniques available in hospitals today.

**Treatment or therapy** is understandably more under the control of the clinician (specialist doctor) than the physicist or technician. Examples are given of a wide range of physical therapies, throughout this book, but their treatment(!) is brief. It is not the purpose of this book to teach you medicine!

# WHAT DO MEDICAL PHYSICISTS DO?

**T**he role of the physicist in medicine is wide-ranging and often crosses scientific and technological disciplines. In the UK the scientific aspects of the work are supported by the Institute of Physics and Engineering in Medicine, the title emphasising the broad nature of the interests. Specialist areas within this can be identified; a particular hospital physics department may be involved in most of these at any one time.

**Medical or clinical engineering** includes the replacement of body parts such as heart valves and joints, and the fitting of powered artificial limbs. There has been a growth in recent years of electronic aids for the disabled, not only to assist in physical actions such as locomotion and eating, but also to produce sophisticated communication aids. The term **bio-engineering**, sometimes used for this work, is really much wider, including engineering in any biological system, overlapping with biotechnology. Chapter 1 deals with mechanical aspects of the body. The manufacture and fitting of hearing aids and lenses for defective vision is not usually included under this heading, being dealt with by specialists such as audiologists, ophthalmologists and opticians. The physics of these aids is covered in Chapters 2 and 3.

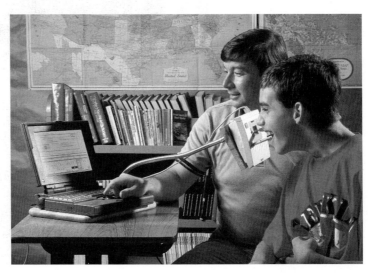

Aiding the communication skills of a physically disabled boy.

**Biomedical measurement and medical electronics** is a rapidly growing field as a result of developments in microelectronics. There is a considerable amount of local cooperation between clinician and physicist or electronic engineer to produce new equipment to meet special requirements. For example the pH in the oesophagus can be monitored by a radio link to the measuring device, to enable surgeons to diagnose problems with the acid contents of the stomach refluxing. Such a project can only succeed with the collaboration of several different specialists. Examples will be found in the chapters of Theme 2. A common role for the physicist is the evaluation of the performance and potential hazards of major equipment, for example a new X-ray scanner.

**Medical ultrasonics**, using echoes of high-frequency sound from the body, has become a very common method for investigating many features of soft tissue, especially in obstetrics and cardiology. The physicist provides technical backup for routine procedures and is involved in the development of new techniques, such as the increased use of computers in the formation of dynamic images. This is covered in Chapter 8.

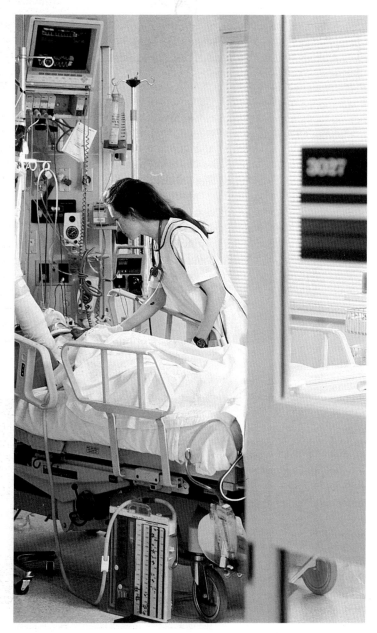

Monitoring the oxygen and carbon dioxide content of exhaled air from a patient in intensive care.

**Radiology** began with the application of X-rays to 'see' into the body, immediately after their discovery by Roentgen, perhaps the most dramatic event in the history of medical physics. The subsequent realisation of the damaging

The hands of the Duke and Duchess of York; radiographs published in the *Illustrated London News*, 18 July 1896, just a few months after Roentgen's discovery was announced.

effects of radiation has reduced the extent of their use, but recent technological advances have re-established an important place for X-rays in the modern imaging department. These changes give considerable scope for the role of the physicist in research and development. These issues are presented in Theme 3.

**Nuclear medicine** is the use of radioisotopes to measure and image body functions and organs. It was born after the development of the atomic bomb in the second world war, as a result of some physicists searching for a beneficial use for artificial radioactivity to 'atone' for the destructive power of the bomb. The use of radioisotopes often complements radiology, and the physicist will be involved in similar tasks such as the production of computer software to provide more clinically useful information. This is described in Chapters 10 and 12. Both types of radiation sources can be used therapeutically at high doses.

**Radiation protection** for both patients and staff from the unintended irradiation by sources in medical use is the responsibility of the physicist. Specialists in radiation physics have an important role in monitoring and evaluating the effects of other radiation exposures, such as those from nuclear reactor accidents. This is dealt with in Chapter 11.

Lateral NMR scan of a man eating a plum.

**Nuclear magnetic resonance** is a completely different imaging technique, developed in the last twenty years. It is ideal for studying complex soft tissues such as the brain. The physics and engineering involved in developing this equipment have been substantial, so that, although expensive to build, NMR scanners are now becoming an integral part of the provision of medical imaging.

**Computing** is now an integral part of much of the routine processing of biomedical measurement, as described in Chapter 4. This is often the responsibility of the physicist, perhaps working with software and electronic engineers in larger hospital departments. These staff may also have a role in the wider implementation of data processing throughout the hospital.

# WHAT ARE THE CAREER OPPORTUNITIES IN MEDICAL PHYSICS?

**T**he development of physics within hospitals has in the past been largely a response to local need. This is reflected, in the UK, in the range of sizes of departments and the variety of functions that they carry out. In the USA and many other countries there is no separate department; physicists and technicians are attached to appropriate medical departments.

Physical scientists may enter the profession directly after graduating (with a good honours degree or equivalent). An increasing number study for a higher degree or enter medical physics with experience of work in another field. Physical science technicians enter with a training qualification such as BTEC, in the UK.

In-service training leading to professional qualifications, is organised in the UK by the Institute of Physics and Engineering in Medicine and the Hospital Physicists Association. There are similar organisations for medical physicists in other countries, from whom further details of the type of work and careers can be obtained.

# HIGH TECH MEDICINE – IS THERE NO ALTERNATIVE?

**Y**ou may gain the impression from reading this book that all the developments in medicine at the present time involve complex expensive equipment; that progress is only to be achieved by the use of 'high' technology. That would certainly be a limited view, for a number of reasons. This book has tried to include current medical developments which involve physics. It does not attempt to describe their relative importance in securing the health of the population. In measurement and diagnosis the stethoscope and sphygmomanometer are much more commonly used than the CT scanner or Gamma camera. There has also been little opportunity in the space available to describe the 'complementary' medical treatments such as hydrotherapy (exercise in water), or osteopathy (manipulation of bones), though the merits of these and other 'low tech' methods are now being increasingly recognised.

There is a crisis in health care in many western countries – the financial demands made of the services continue to increase with no indication that these will ever be fully met. In less prosperous countries the situation is of course even more serious. In these circumstances high tech medicine becomes something that will always have to be rationed and the difficult question is 'Who decides who gets what treatment?'

In the UK one approach being developed to try to answer this is QUALY, quality adjusted life years. This takes into account the average additional years gained from a treatment, multiplied by a quality factor. On this basis it is less of a priority to treat cancer and heart disease (since the sufferers are usually the elderly), and much more important to ensure good maternity services, child health and in adults attend to conditions such as varicose veins and hip replacements.

# QUESTIONS FOR DISCUSSION

In *High Technology Medicine, benefits and burdens*, Bryan Jennett, an experienced neurosurgeon, teacher and administrator reviews these issues and includes the following list of inappropriate uses of high technology:

**Unnecessary** – because the desired objective can be achieved by simpler means.

**Unsuccessful** – because the patient has a condition too advanced to respond to treatment.

**Unsafe** – because the complications outweigh the probable benefits.

**Unkind** – because the quality of life after rescue is not good enough, or its duration for long enough, to have justified the intervention.

**Unwise** – because it diverts resources from activities that would yield greater benefits to other unknown patients.

What examples can you think of from your own experience, that might come into each of these categories? Do all of the group discussing this agree or do different personal experiences affect our views?

How would the QUALY approach compare with this?

How would a health service decide what is appropriate?

## *Further reading*

Institute of Physics and Engineering in Medicine
*Physical sciences in medicine* – An illustrated booklet of career information on medical physics.
*Website* http://www.ipem.org.uk

# BODY PHYSICS

**H**ow well do you know your own body? Well enough to know its capabilities and limitations, perhaps. You may have been helped in this by your previous studies in science, linking personal experience with objective knowledge. You will also know something of the complexity of the body, and the range of sciences which are used in studying its function. In this theme we will study those parts of the body where physics has a substantial contribution to make in answering vital questions such as:

Why do so many people suffer from bad backs?

What are a person's daily energy needs?

Why can we not see colours when light intensities are low?

How does loud noise damage our hearing?

Will the vaulter's pole absorb his kinetic energy supply? ■

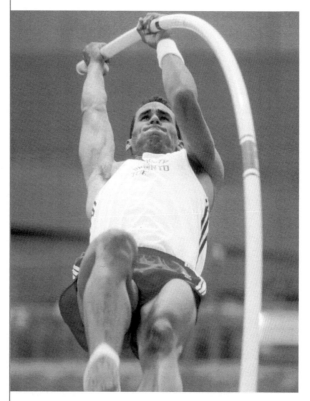

Energy transfer during pole vaulting.

## PREREQUISITES

Before starting this theme you should have some familiarity with the following:

● The basic structure of the human body.
● The concepts of force, work, energy, power, temperature and waves.
● The principle of moments and the resolution of forces.
● The nature of light and of sound.

(You will need to refer to GCSE or A/AS-level texts at some points if these are unfamiliar to you – or forgotten!)

# 1

# HUMAN MECHANICS

The balance of the gymnast and the strength of the weightlifter (Fig 1.1) are two examples of the body's abilities. You may envy their achievements or be thankful that you don't have their bodies! This chapter may not enable you to gain the former or avoid the latter, but you should learn how the body functions in both these and more everyday situations.

## LEARNING OBJECTIVES

*After studying this chapter you should be able to:*

① describe how muscles and bones operate to produce body movements;

② give examples of levers in the muscular-skeletal system of the body;

③ provide a simple analysis of:
   (a) the forces involved in standing, bending and lifting
   (b) the interactions of the body with the ground in walking and running;

④ estimate the power typically provided by muscle;

⑤ describe how the body maintains a constant temperature, and give examples of how this may be assisted;

⑥ recall typical human energy needs and energy values of typical food types.

**FIG 1.1** The body balanced, and the body strong.

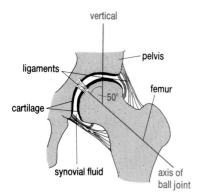

**FIG 1.2** Structure of the hip joint.

**FIG 1.3** Pairs of muscles raising a leg.

## Muscles and skeleton

The human body is supported by a framework of some 200 bones connected together to make the skeleton. Bone is a living material; stiff, light and strong. Individual bones have the optimum properties required for their particular function. For example the femur, which must withstand forces of up to thirty times body weight, is a thick-walled, hollow cylinder with internal cross struts. This gives an excellent strength-to-weight ratio. The skeleton is held together at joints by ligaments. There are several kinds of joints to allow the considerable flexibility of human movement. These range from the plane joints of the toes through pivot (neck), hinge (elbow), ellipsoid (wrist) and saddle (ankle), to the ball and socket of the shoulder and hip which gives maximum freedom of movement (Figure 1.2). The illustration shows features common to all joints, namely the layers of the tough cartilage and the interposed synovial fluid, which prevent bone damage during movement.

Movement of the skeleton is effected by the muscles, which are attached to the bones by tendons. Muscle is a flexible fibrous material which surrounds the skeleton. (You can see the fibres in the 'meat' of a butcher's joint such as a leg of lamb.) Movement is produced by a tensing of the muscle which causes it to become shorter and fatter and pull on the bone. To restore the bone to its original position there is always an opposing muscle acting in antagonism. Hence to raise the leg as shown in Figure 1.3 the rear calf muscle and the front thigh muscles contract whilst the opposing muscles relax.

### Some body figures

| | |
|---|---:|
| number of bones in the skeleton | 206 |
| number of joints in the skeleton | 187 |
| types of joints | 6 |
| number of voluntary muscles | 620 |
| (the involuntary muscles are those of the heart and the intestines) | |
| percentage of body weight which is bone | ~17 |
| percentage of body weight which is muscle | ~40 |
| strength of bone compared to reinforced concrete | ×4 |

## Forces and levers

The tensed muscle exerts a force on the bone at the place where it is attached by the tendon. This produces a turning effect about the joint. The system acts as a lever with the joint as fulcrum. The muscle provides the effort ($E$) and the load ($L$) may be the weight of the part of the body being moved, or may include an external load. From the principle of moments for a body in equilibrium:

$$L \times \text{distance of load from fulcrum} = E \times \text{distance of effort from fulcrum}$$

Often the effort is applied closer to the fulcrum than the load, giving the system a mechanical advantage of less than one, since:

$$\text{mechanical advantage} = \frac{\text{load } (L)}{\text{effort } (E)} = \frac{\text{distance of } E \text{ from fulcrum}}{\text{distance of } L \text{ from fulcrum}}$$

(a)

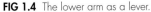

(b)

**FIG 1.4** The lower arm as a lever.

This arrangement, illustrated in Figure 1.4 gives a greater control to the body movement, a small movement of the biceps muscle, with a large force, produces a much larger movement of the book.

### QUESTION

**1.1**    **(a)** The head can be tipped back by the pull of the muscles on the back of the neck, with the top of the spine, the atlas vertebra, as fulcrum. Draw a simple diagram of this lever, showing the forces acting.

       **(b)** Draw a similar diagram for the action of the calf muscle when a person stands on tiptoe, pivoting on the ball of the foot.

## Solving lever problems

We can work out the forces in muscles when they are producing movements, by considering only the part of the body involved. Look at Figure 1.4, where we can take moments about the joint as the fulcrum. In this treatment it is assumed that the force of the humerus on the lower arm acts through the fulcrum, at a single point in the joint and that the tendon is acting at a single point on the lower arm. When the forearm is bent at right angles, with the dimensions shown in Figure 1.4(b), the contractile force of the biceps ($E$), and the reaction of the upper arm, can be calculated by taking moments about F.

$$E \times 50 \text{ mm} = (20 \text{ N} \times 150 \text{ mm}) + (5 \text{ N} \times 400 \text{ mm})$$
$$50E = 3000 \text{ N} + 2000 \text{ N}$$
$$E = 100 \text{ N}$$

Resolving forces vertically:

$$E = R + 20 \text{ N} + 5 \text{ N}$$
$$R = 75 \text{ N}$$

### QUESTION

**1.2**    A champion weightlifter can raise a mass of 250 kg. His forearm has a length from elbow joint to palm of hand of 0.5 m, and a weight of 30 N. Assuming that the force of the biceps muscle acts at a point 0.07 m from the fulcrum, calculate the maximum values of the force in each biceps and the reaction in each upper arm.
(Take $g$, the gravitational field strength as 10 N kg$^{-1}$)

## The vertebral column

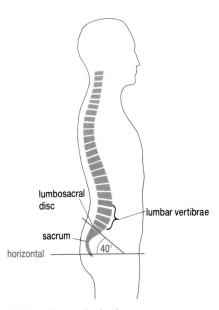

**FIG 1.5** The vertebral column.

The backbone provides the main support to the body, and is required to make a great variety of movements. It consists of 33 vertebrae, of which nine are fused together to make up the sacrum and the coccyx, and the remaining 24 are separate, covered with cartilage and interspersed with tough fibrous pads called discs. It is the discs which permit the free bending of the spine, whilst cushioning the bones from the compression of body weight and protecting the spinal cord which runs down the centre of the vertebrae. The discs are subjected to considerable wear throughout life and they also lose some of their toughness with age. If the stresses become too large a disc can slip out of position. It is not therefore surprising that about 80 per cent of people suffer from back pain at some time in their lives. The lowest part, the lumbosacral region is the most susceptible. Because, as Figure 1.5 shows, it is here that the normal backbone deviates most from the vertical. The sacrum is attached firmly to the pelvis so has much more limited movement, whereas the lumbar vertebrae are connected to each other only through the discs.

## 1.2 STANDING, BENDING AND LIFTING

### Standing

In normal posture the lumbosacral disc lies at an angle of about 40° to the horizontal; this is called the lumbar lordosis. The weight of the body on this disc is about 60 per cent of the total body weight, $W$. It is called the superincumbent weight, (the weight lying on). This will be balanced by the support force of the sacrum, $S$. These forces can be resolved to give a **compressive stress** (squashing), perpendicular to the disc, and a **shear stress** (twisting) parallel to the bone surface (Figure 1.6).

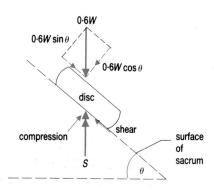

**FIG 1.6** Forces on the lumbosacral disc.

compressive stress = $S \cos \theta$
shear stress = $S \sin \theta$

For normal posture

compressive stress = $0.6W \times \cos 40° = 0.46W$
shear stress = $0.6W \times \sin 40° = 0.39W$

The discs are designed to withstand compression rather than shear, so this shear stress is potentially damaging. The situation is made worse if a person has poor posture so that the lumbar lordosis is greater than 40°, which may be caused by weakened muscles, for example as a result of pregnancy.

When we stand on two legs our body weight is equally divided between each leg. At the head of the femur, where the leg connects to the body, the superincumbent weight is about $0.7W$. The axis of the ball joint is usually at an angle of about 50° to the vertical (Figure 1.2).

When the body is supported on one leg the body's centre of gravity has to shift so that it is over the supporting foot. The superincumbent weight then has a moment about the femoral head, which increases the total force on it to about 2½ times body weight (compared with just over ⅓ body weight when standing on two legs).

---

**QUESTION**

**1.3** Resolve the force at the femoral joint to find the compressive and shear stresses on this joint when a person of mass 70 kg stands on two legs. How does this compare with the stress on the lumbosacral disc?

---

### Bending

Damage to the back occurs most commonly during bending and lifting loads. A simple analysis of the forces involved shows why this is so. A range of muscles, the erector spinae, link the vertebrae of the spine to the sacrum and ileum of the pelvis.

We can simplify the arrangement by considering the spine as a rigid column, pivoting on the sacrum at the lumbosacral joint (Figure 1.7). The superincumbent weight, $W_b$, of the trunk, head and arms acts at their combined centre of gravity, G, which is about two-thirds the way up the trunk. This is balanced by the lift provided by the muscles to the upper part of the body, which can be represented as a single force through G. X-ray observations indicate that this force is applied at about 10° to the vertebral column.

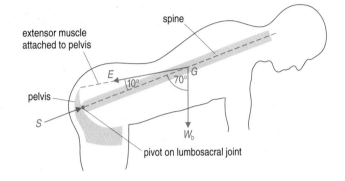

**FIG 1.7** Bending.

Let us now calculate the tensile force $E$, of the muscles and the compressive force on the lumbosacral disc, $S$, when the body is bent at an angle of 70° to the vertical (Figure 1.7). The three forces $W_b$, $E$ and $S$ are in equilibrium, so the line of action of $S$ is along the column, so that it passes through G.

Average body weight for a person is about 700 N, so

$$W_b = 0.6\,W = 420\text{ N}$$

Resolving forces perpendicular to the column,

$$E \sin 10° = W_b \sin 70°$$
$$E = 2275\text{ N}$$

A considerably greater force than body weight because of the acute angle of the muscles' action.

Resolving forces parallel to the column,

$$S = E \cos 10° + W_b \cos 70°$$
$$S = 2385\text{ N}$$

This substantial force reduces the thickness of the disc by about 20 per cent.

## Lifting

We can extend this treatment to consider the forces involved when bending and lifting an object, for example a chair of weight 300 N. Figure 1.8 shows the forces involved, with the load acting at the top of the trunk in line with the shoulder joint (the distance $L$ from the pivot). There are now four forces acting and so $S$ acts at a small angle to the axis of the column. There is a compressive component $S_c$ and a shear component $S_p$.

Taking moments about the fulcrum, F:

$$E \times 0.67L \sin 10° = W_b \times 0.67L \sin 70° + 300\text{ N} \times L \sin 70°$$
$$E = 4711\text{ N}$$

Resolving forces parallel to the column,

$$S_c = E \cos 10° + W_b \cos 70° + 300 \text{ N} \cos 70°$$
$$S_c = 4886 \text{ N}$$

Resolving forces perpendicular to the column,

$$S_p + W_b \sin 70° + 300 \text{ N} \sin 70° = E \sin 10°$$
$$S_p = 141 \text{ N}$$

The total reaction is given by:

$$S^2 = S_c^2 + S_p^2$$
$$S = 4888 \text{ N}$$

This acts at an angle $\theta$ to the column, where:
$$\tan \theta = S_p/S_c$$
$$\theta = 1.5°$$

**FIG 1.8** Lifting.

## The disc under pressure

A normal lumbar vertebral disc will contract elastically under compression up to a force of about 1000 N. Beyond this it deforms plastically and will rupture when it has contracted by about 35 per cent, at a load of about 15 000 N. Rupture causes the fluid in the disc to be extruded and cause pressure on adjacent nerves in the spinal cord (Figure 1.9). This is what causes the intense pain and muscle spasm associated with disc trouble. Treatments include rest, strengthening the muscles by exercise, skilled manipulation and surgery.

**FIG 1.9** Computed tomography scan showing damaged disc pressing on spinal cord.

> ### QUESTION
>
> **1.4**    **(a)** From the above figures, work out the ratio of the compressive stresses in the lumbosacral disc, for
>      **(i)** standing
>      **(ii)** bending
>      **(iii)** lifting a weight of 300 N
>    **(b)** Explain how a vertical posture when standing and when lifting objects helps to avoid a slipped or ruptured disc.

## Ground forces

A body standing on the ground is supported by a force equal and opposite to the body's weight. This force $V$, will therefore have a magnitude of $W$ and act vertically upwards. When the body is moved relative to the ground the force will change. When the foot lands on the ground at the end of a step, a force is needed to decelerate it to rest. When the foot is thrust off the ground for the next step an accelerating force is required. In each case the ground force will be greater than $W$, and will be the resultant of a vertical force $V$, and friction $F$, as illustrated in Figure 1.10. The forces will of course increase as the speed with which the foot meets and leaves the ground increases.

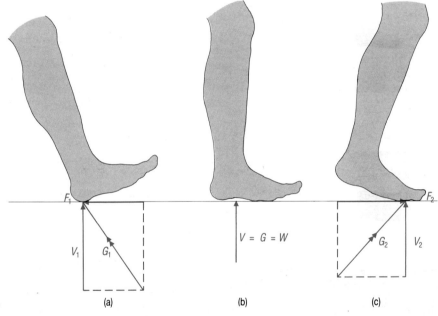

**FIG 1.10** Ground forces while walking and running.

The maximum size of the vertical force depends on the material of the ground. If this is soft it may undergo considerable deformation in order to provide the force needed. The maximum size of the frictional force depends on the nature of the surfaces and the size of the vertical force. This is expressed as,

$$F = \mu V$$

where $\mu$ is the coefficient of static friction between two surfaces. Thus if the horizontal force needed is greater than $\mu V$ the foot will slip. Typical values of $\mu$ for shoes on ground are 0.6 to 0.75.

The resultant force of the ground $G$, has magnitude given by:

$$G^2 = V^2 = F^2$$

and direction given by:

$$\tan \theta = F/V$$

and the horizontal forces required for walking are about 15 per cent to 20 per cent of body weight.

> **QUESTION**
>
> **1.5** **(a)** With reference to the previous section explain:
> **(i)** why slipping does not normally occur while walking;
> **(ii)** why slipping is more likely to occur when the stride is long and when $\theta$ is small.
> **(b)** Imagine you have to cross a wet and muddy ploughed field on a cross-country run. Use the above account to explain why your options are to slip across the surface slowly or plough through, ankle deep, with a great effort.

## Walking

The forward acceleration of each leg requires a horizontal force given by:

Force = mass × acceleration

There is also a rotation of the leg as it moves forward, in which the turning force or torque produces an angular acceleration. The relationship here is given by:

torque = moment of inertia × angular acceleration

where the moment of inertia depends both on the total mass of the leg and its distribution. Thus any change such as the wearing of heavy boots, will result in extra energy expenditure and possibly muscle fatigue.

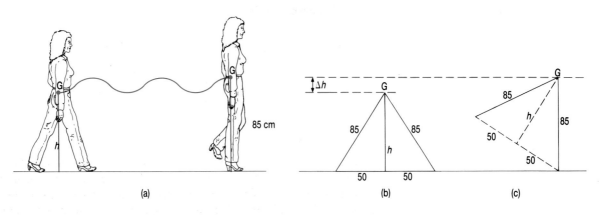

**FIG 1.11** Changing the height of centre of gravity while walking.

Deceleration requires a similar application of force and torque, though this is assisted by gravity. A similar forward swinging movement occurs at the knee. Work is also done in raising and lowering the centre of gravity G, during each step. This is shown in Figure 1.11 which represents a walker of leg length 85 cm taking a stride of 1 m.

The rise in G is

$$\Delta h = 85 - h = 85 - \sqrt{85^2 - 50^2}$$
$$= 16.25 \text{ cm}$$

This is an over estimate as we actually bend our knees at the position where G is highest, which reduces the rise to about half of this value.

# 1.4 ENERGY EXPENDITURE

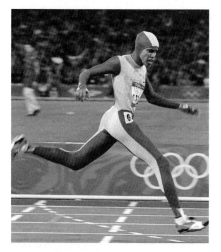

**FIG 1.12** Kathy Freeman conserves angular momentum while running in the Olympic Games.

## Running

Accelerations and torques in running are much greater, in order to produce the higher speed. A fast sprint (about $9 \, \text{m s}^{-1}$) is six times faster than a typical walking speed. This is achieved by a combination of an increase in stride frequency and in stride length. To offset the considerable angular acceleration of the swinging leg, the opposite arm is swung forward. This causes the upper and lower parts of the body to twist in opposite directions to conserve angular momentum (Figure 1.12).

Energy is rather like wealth – a description of a situation or condition, rather than a thing in itself. We are usually interested in changes of energy and of wealth, not absolute values, and the currency we use is the joule. The human body expends energy doing mechanical work and maintaining a constant body temperature. The income is from the food we eat. Excess of income over expenditure is stored in the body, mainly as fat. Excess of expenditure over income leads to fatigue, ill health and eventually the possibility of death.

## Doing work

Work is done by the body in moving against gravity, external forces of friction and air or water resistance, and internal resistance of the body. In walking for example, the raising of the centre of gravity with each step, as described in the previous section, results in an energy expenditure of:

$$\text{work done per stride} = \text{weight} \times \text{rise}$$
$$E = mg\Delta h$$
$$= 70 \, \text{kg} \times 10 \, \text{N kg}^{-1} \times 0.08 \, \text{m}$$
$$= 56 \, \text{J (joules)}$$
$$\text{work done per km} = 56 \, \text{kJ (for strides of 1 m)}$$

Assuming a walking speed of $5 \, \text{km h}^{-1}$

$$\text{power} = 5 \, \text{km h}^{-1} \times 56 \, 000 \, \text{J} / 3600 \, \text{sh}^{-1}$$
$$P = 78 \, \text{W (watts)}$$

In jumping, the muscles do work to produce kinetic energy ($E_k = \frac{1}{2}mv^2$) at take off. This is converted to potential energy ($E_p = mgh$) as the body is propelled upwards. In high jumping and pole-vaulting the vertical height of the centre of gravity is all-important. In long jump and triple jump, speed and height are both important in carrying the jumper forward.

## INVESTIGATION

### The mechanics of jumping

This is an opportunity to measure performance and study how it may be improved (in yourself or a suitable volunteer!).

### Standing jump

1. Measure take-off speed $u$, when jumping vertically. The normal laboratory ticker-timer will probably not be suitable for this. You could devise a photographic method, but a simpler way is to measure the height of $\Delta h$ jumped (Figure 1.13) and use the equation

$$v^2 = u^2 + 2gh$$

For example, a good jump would be 1.0 m, giving a take-off speed of $4.43 \, \text{m s}^{-1}$.

### Long jump

A simple model of long jumping is that the athlete runs as fast as possible and jumps vertically at the take-off point.

FIG 1.13 Measuring jump height with chalk marks.

2. Measure maximum sprint speed $v_{max}$ and length of long jump $s$.

3. Check whether this simple model is appropriate for your jump, using the equations for a projectile:

$$h = ut + \tfrac{1}{2}gt^2$$
$$s = v_{max}t$$

For example, a top athlete such as Carl Lewis can run at approximately 10 m s$^{-1}$. Assuming his vertical take-off speed is that of a 'good' jumper as in **1.** above, then:

$$0 = 4.43t - 4.9t^2$$
$$t = 0.904 \text{ s}$$
$$s = 9.04 \text{ m}.$$

Lewis won the 1996 Olympic gold medal with a jump of 8.50 m, so this simple model seems to give a reasonable prediction for a top athlete. How do your results compare?

## High jump

4. By considering the conversion of $E_k$ to $E_p$, derive an expression to show that the height jumped $\Delta h$ depends on the square of the take-off velocity.

   The height jumped is the distance that the centre of gravity is moved upwards. For a standing adult the distance of the centre of gravity from the ground $H_c$ is about 1 m. Assuming it does not rise much above the bar, we can calculate the expected jump height from this expression. For example, with a take-off velocity for a standing start of 4.43 m s$^{-1}$, $\Delta h$ is 1 m, giving a bar clearance $H$ of 2 m. The women's Olympic record (by Louise Ritter in 1988) is 2.03 m, so you can see that horizontal running speed has little contribution to make.

5. Calculate the height of your centre of gravity above the ground (about 55 per cent of full height for women, 57 per cent for men). Measure the bar clearance height at various take-off speeds. Compare the values of $\Delta h$ calculated from $H - H_c$ with the predictions from $H = u^2/2g$. For the successful high jumper technique and body position are much more important than sprint speed (Figure 1.14).

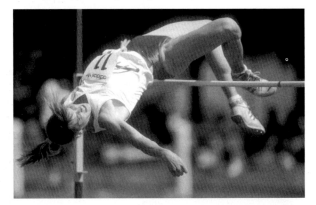

FIG 1.14 High jumper keeping her centre of gravity as low as possible.

---

QUESTION

1.6 With reference to the previous investigation explain:
(a) why good long jumpers are fast sprinters,
(b) why good high jumpers are tall,
(c) the function of aerial leg kicks in long jumping,
(d) the function of the diving or flopping technique in high jumping.

# Basal metabolic rate

The body is continuously performing involuntary functions which require energy. These include the beating of the heart, breathing, digestion and growth. This maintenance expenditure of energy is called the basal metabolic rate (BMR). In children BMR is high because of the energy used in the biosynthesis of components for growth. The rate levels off as we mature, falling again in old age as all processes slow down (Figure 1.15). From childhood onwards, males have a higher BMR than females, on average. This is because of the lower proportion of fat in the male body. If this is excluded from the calculations, both sexes have similar BMRs.

The energy expenditure can be calculated indirectly from temperature or respiration measurements as described below. Alternatively we could calculate the work done in the various processes. In Chapter 6 for example, the power of the heart is calculated from a knowledge of its volume, pulse rate and the pressure of the blood leaving it.

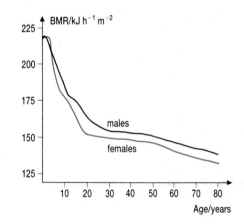

**FIG 1.15** Basal metabolic rate variation with age and sex.

# Getting warm and keeping cool

The values shown in the graph (Figure 1.15) are averages, individuals show a considerable variation. One of the factors which causes this variation is the maintenance of constant body temperature. The human body operates best at around 37 °C, and will cease to function if the temperature changes by more than about 6 °C from this, as described in Table 4.2 in Chapter 4. In temperate climates, such as Britain, the body will need to expend energy to keep itself at a higher temperature than the surroundings. In tropical climates, or during periods of intense exercise, the body will need to expend energy to cool itself, like a refrigerator. So the amount of energy expenditure will depend on the ambient temperature, the work being done by the body, the surface area of the body and what clothes are worn.

The body can lose energy by conduction, convection, radiation, evaporation (of perspiration), respiration (breathing out warm air and water vapour) and excretion. The relative importance of each depends on the condition of the body and its surroundings. Table 1.1 summarises the processes and gives some typical values.

The amount of energy released in body processes depends mainly on the **volume** $V$, of the tissues. The rate of loss of this energy depends on the **surface area** $A$, of the body. The smaller the body, the higher is the **surface to volume ratio**, $A/V$. This means that for example babies are much more at risk of hypothermia (low temperature), than adults. It also follows that smaller people need to eat more, relative to their mass, than larger people. Typical values of these quantities are given in Table 1.2.

**TABLE 1.1** Energy loss processes

| PROCESS | DETERMINING FACTOR | VALUES |
|---|---|---|
| conduction | temperature difference, area of surface, conductivity | very low due to insulating properties of fat, hair, clothing (skin 0.042 W m$^{-1}$ K$^{-1}$) |
| convection | temperature difference air speed around body | low, except in draughts |
| radiation | absolute temperature, area and nature of surface (see Chapter 4 for details) | 100 W for an unclothed body at 22 °C |
| evaporation | temperature difference, area of skin exposed, humidity, air movement | 625 W maximum |
| respiration | air temperature and humidity | 20 W resting at 22 °C |

| TYPICAL VALUES | STUDYING AT 22 °C | SUNBATHING AT 32 °C | WALKING AT −18 °C |
|---|---|---|---|
| % of body clothed | 85 | 15 | 95 |
| heat loss rate /W | 170 | 400 | 400 |
| % loss by: | | | |
| 1  radiation | 21 | 8 | 8 |
| 2  conduction and convection | 67 | 10 | 50 |
| 3  evaporation | 10 | 80 | 2 |
| 4  respiration and excretion | 2 | 2 | 40 |

**TABLE 1.2** Body sizes and surface area to volume ratios

| | SURFACE AREA $A$/m$^2$ | VOLUME $V$/dm$^3$ | SURFACE AREA TO VOLUME RATIO $\frac{A}{V}$/m$^2$ dm$^{-3}$ |
|---|---|---|---|
| Smaller person | 1.53 | 56 | 0.0273 |
| Larger person | 1.97 | 80 | 0.0246 |

## QUESTIONS

**1.7**   With reference to the data in Table 1.1, answer the following:
  **(a)** The specific latent heat of sweat is about 2425 kJ kg$^{-1}$. What is the rate of its excretion from the skin to produce the maximum rate of energy loss quoted?
  **(b)** Why is the percentage of energy loss by evaporation (i) high when sunbathing and (ii) low when walking in cold conditions?
  **(c)** Why is a higher proportion of the energy loss by conduction and convection, when a person is studying?

**1.8**   Because of the importance of surface area in determining BMR, the values are expressed in Figure 1.15 as kJ h$^{-1}$ *per square metre*. From the figure and Table 1.2, calculate the range of power dissipation, in watts, for males and females, in their seventeenth year.

## Total energy requirement

This is the BMR value added to the work done in activities. It can be measured directly by putting the subject in an isolated room, (called an Atwater chamber, after its designer), and monitoring all energy exchanges. For example temperature changes are measured by water circulating in pipes through the chamber. Any work done which may be stored (such as raising weights), is added to this. Work which is not stored, (such as exercise movements) will produce a temperature rise so need not be separately considered. Typical ranges of values are given in Table 1.3. The lower values in the range are for young or old subjects, for females and for less vigorous activity. The highest values are usually for heavier fit adult males – and even fitter females!

These figures lead to estimates of total energy expenditures of about 10 MJ for an average woman, between the ages of 15 and 50, with a moderately active lifestyle and about 13 MJ for a man of similar age and lifestyle. The differences are mainly due to different BMR and body weights. A woman in the latter half of her pregnancy, and while lactating to feed her baby, requires an extra 1.5 MJ.

**TABLE 1.3** Energy conversion

| ACTIVITY | CONVERSION RATE/W |
| --- | --- |
| Resting (BMR) | 60–100 |
| Sitting or standing still | 80–150 |
| Moving around classroom or home | 100–300 |
| Walking or 'light' sports, housework | 150–450 |
| Moderate sports, e.g. swimming, gymnastics | 300–550 |
| Climbing stairs, carrying heavy loads | 400–850 |
| 'Heavy' sports, e.g. running, rowing, squash | 400–1400 |

## QUESTION

**1.9** From the data in Table 1.3 calculate the energy expenditure for yourself, on an active and on an inactive day. How do your answers compare with the totals given above?

## 1.5 ENERGY INCOME

### Input

The body's energy is supplied from the chemical energy of the food we eat.

There are three types of fuel foods: carbohydrates, fats and proteins. The digestive system breaks these substances down into chemically simpler ones: (a) carbohydrates to sugars such as glucose, (b) fats (triglycerides) to glycerol and fatty acids, and (c) proteins to amino acids. These simpler molecules are absorbed into the bloodstream and transferred to the cells, where they are oxidised, by the oxygen inhaled by the respiratory system. Enzymes are required to catalyse the reaction, but, as when any fuel is burnt, energy is released. The process of releasing energy by oxidation in the body is called **aerobic combustion** or **respiration** and its general reaction is:

fuel + oxygen = carbon dioxide + water + energy

and the particular equation for glucose is:

$$C_6H_{12}O_6 + 6O_2 = 6CO_2 + 6H_2O + 4 \times 10^{-18} J.$$

The respiration reaction is the reverse of photosynthesis in which plants 'fix' energy from sunlight. The energy released by combustion can be measured by calorimetry, typical values (of standard enthalpies) are given in Table 1.4. In the body the amount of energy released will normally be less than this, due to incomplete digestion of the foods and incomplete absorption of substances into the blood, giving the available energy values shown.

Thus the energy used at any time can be measured by the consumption of oxygen. A **spirometer** is used to monitor the conversion of oxygen to carbon dioxide. The average energy liberated from the combustion of carbohydrates, fats and proteins is 20.17 kJ per litre of oxygen. The rate of oxygen intake is a useful measurement to make on people who are in training for sports or under treatment for obesity.

An alternative breakdown of glucose to lactic acid can also release energy. This does not require oxygen, it is **anaerobic respiration**. It occurs therefore when energy is required in the absence of oxygen, for example during extreme exercise. The process is inefficient, liberating only about one twentieth of the energy released by the aerobic process. It also causes feelings of fatigue and pain as lactic acid is poisonous. It is this which sets a limit on an athlete's performance. Recovery consists of the oxidation of the lactic acid and the making up of the oxygen shortage in the muscles.

The food normally used as a fuel is a carbohydrate, with fat being drawn on as a reserve, protein being used only in extreme cases of starvation or overeating. In the former the protein is not then available for growth or maintenance of the body. In the latter the body needs to remove the excess consumption. Usually excess fuel food can be stored in the body.

**TABLE 1.4** Foods as fuels

|  | STANDARD ENTHALPY OF COMBUSTION/MJ kg$^{-1}$ | AVAILABLE ENERGY/MJ kg$^{-1}$ |
|---|---|---|
| carbohydrate | 17.2 | 16.5 |
| fat | 39.4 | 37.5 |
| protein | 23.4 | 16.5 |

## QUESTIONS

**1.10** What is the BMR of a person who, when resting, consumes 1.5 dm$^3$ oxygen in 5 minutes? Give your answer in **(a)** kJ h$^{-1}$ **(b)** W.

**1.11** Explain how an increase in pulse rate and panting (deep, fast breathing), helps an athlete perform better. If an athlete is fatigued, how does massaging the muscles help?

## *Storage*

**Fat** is the major energy store of the human body. Typically 12 per cent of a man's weight and 25 per cent of a woman's weight is made up of two types of fat, to be drawn on when energy is required. Fat is stored in adipose tissue, distributed throughout the body, especially under the skin where it provides good insulation. Adipose tissue is the white fat, and more is found in overweight people. Brown fat is usually found elsewhere in the body, and particularly in lighter people. As the brown fat has a higher metabolic rate it is thought that this can provide a less bulky energy store. Table 1.4 shows that fat provides double the energy produced by carbohydrate for a given mass. **Carbohydrate** breakdown products are stored in different parts of the body. Glycogen can be stored in the muscles or the liver, where it is broken down by enzymes to glucose, when required. It can also be converted to the high energy store substance adenosine triphosphate (ATP), which is again held in the muscle. The major supply of energy from carbohydrates is the glucose in the bloodstream.

Excess intake of food will result in excess body weight, or obesity. This is known to cause many common ailments; physical such as heart, circulatory, and respiratory malfunction, and psychological, for example poor self image. Despite the impression you may have gained from the above sections, that it seems a simple matter for a person to decide what energy intake is required, obesity is a major problem in Western societies, as witnessed by the size of the 'slimming industry'. This consequence of Western affluence is thrown into sharper relief, when seen against the widespread malnutrition and frequent starvation experienced by people in poorer parts of the world.

Research into the causes and prevention of obesity receives considerably less funding than the profits generated by the UK sales of slimming aids and advice.

## SUMMARY ASSIGNMENTS

**1.12** Obtain details of a current slimming programme which gives some information about energy inputs and outputs. Compose a short appraisal of its likely effectiveness from what you have learnt in this chapter.

**1.13** Hypothermia may be combatted by vigorous exercise, external heating of the body, insulation such as duvets or thermal underwear, and reflecting blankets. Explain how each is appropriate to an at risk group such as elderly people, mountaineers or exhausted runners.

**1.14** This question is about the action of levers in the human body.
  **(a)** The head may be considered as supported on a lever in which the atlas vertebra at the top of the spine acts as pivot.
  **(i)** When the head is upright the pull of the muscles at the back of the neck supports the weight of the head. Copy Figure. 1.16 and draw arrows to represent these two forces acting on the head. Mark the position of the pivot on the diagram.
  **(ii)** The weight of the head is 75 N and its line of action is 0.04 m in front of the atlas vertebra. Calculate the force exerted by the neck muscles which act vertically 0.055 m behind.

**FIG 1.16**

  **(b)** Sketch two diagrams to show examples of two other levers which act in the body. In each case, state the force which causes the load and how the body supplies the effort. On your sketches label load, effort and pivot.
  **(c)** In the type of lever of which the horizontal forearm is an example, the mechanical advantage (MA) is always less than 1. Explain why this is so.

*(O & C June 1997)*

**1.15 (a)** When a person is standing upright and still, the load on the head of one femur (thigh bone) is vertical. The neck of the femur makes an angle of 50° to the vertical as shown in Figure 1.17

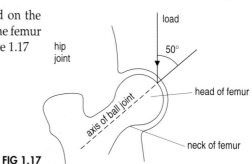

**FIG 1.17**

The weight, *W*, of the person is 700 N and the load on the head of one femur is 0.35*W*. This load will produce a compressive force along the axis of the ball joint and a shear force normal to this axis.
  **(i)** Indicate on your diagram how the compressive and shear components of the load act on the head of the femur.
  **(ii)** Calculate the values of these two components.

**(b)** Figure 1.18 shows the forces acting on the lumbosacral disc of the same person still standing upright. The vertical downwards load, 0.60W, and the upwards support force, S, are equal in magnitude but produce compressive shear forces on the disc.

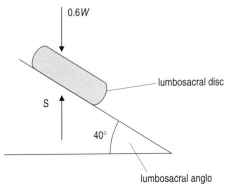

**FIG 1.18**

**(i)** If the effective area of the disc which supports the compressive load is $3.0 \times 10^{-4}$ m², calculate the compressive pressure on the disc.

**(ii)** A person with a poor posture may curve the spine and change the angles of the vertebrae relative to each other. Describe the consequences of such poor posture over many years for the lumbosacral disc and for the person.

*(NEAB June 1996)*

**1.16 (a)** Figure 1.19 shows the variation with age and gender of the basal metabolic rate (BMR).

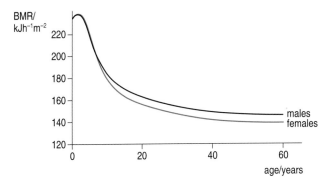

**FIG 1.19**

**(i)** Explain the term *basal metabolic rate*.

**(ii)** Explain the importance of surface area with respect to body volume for the daily energy requirement of a person.

**(b)** Use Figure 1.19 to estimate the rate of energy transfer through the surface of a 20-year-old male of surface area 1.7 m².

**(c)** The average daily energy input for this male is 16 MJ.

**(i)** What fraction of this total energy input is required to sustain metabolic activities?

**(ii)** Assume that the remainder of the 16 MJ is converted in muscle activity with an efficiency of 25%, and that all of the energy required for metabolic processes ends up as heat energy. Calculate an average rate of heat energy production due to muscle activity during the 24-hour period.

**(iii)** The actual rate of heat production during muscle activity is about 300 W. Explain why your answer to **(ii)** is different from this value.

*(UCLES Feb 1997)*

**1.17** An athlete of mass 80 kg completes a 1500 m race in a time of 4.0 minutes. For each stride during the race, the centre of mass of the athlete rises and falls through a vertical distance of 15 cm. The average length of each stride is 1.8 m.

**(a)** Show that the average rate of working against gravity during the race is about 400 W.

**(b)** Suggest why the total useful power output of the athlete's muscles during the race may be different from that in **(a)**.

**(c)** The total output power for this athlete is 1600 W. The muscles are working with an efficiency of 25% and the remaining 75% is converted into thermal energy.

    **(i)** Calculate the rate of production of thermal energy for this athlete during the race.

    **(ii)** Perspiration and its evaporation remove 40% of the thermal energy generated in the muscles during the race. 2300 J of thermal energy are required for the evaporation of 1 g of sweat. Calculate the mass of sweat evaporated from the athlete as a result of the race.

**(d)** Explain why, at the end of marathon races, athletes are sometimes wrapped in foil.

*(OCR June 1999)*

# THE EYE AND VISION

The human body is sensitive to most electromagnetic radiations. Microwave and infrared produce the sensation of warmth. Ultraviolet and ionising radiations can produce chemical changes and cause biological damage such as skin cancer. The eye is the one organ of the body designed to respond specifically, to receive (or perceive) visible light, a part of the electromagnetic spectrum having wavelengths between about 380 and 760 nanometres. Of all the sense organs the eye has the greatest number of nerve cells dedicated to it, emphasising the importance of vision in our lives. There are two parts to the eye's perception; detection by the sensitive cells of the retina and image formation by the optical system. In this chapter we will consider how each of these work, and how to correct them when they do not.

## LEARNING OBJECTIVES

*After studying this chapter you should be able to:*

① give an outline description of the structure and function of the eye;

② explain how the retina responds to light, including the visual effects of colour, acuity or resolution, adaptation to illumination levels and persistence;

③ give examples of the social implications of colour perception

④ explain how the optical system produces images, including the optical features of depth of focus, depth perception and resolution (diffraction);

⑤ explain how optical defects of vision may be corrected with lenses.

## 2.1 STRUCTURE AND FUNCTION OF THE EYE

The eye is often referred to as the living camera. If we consider the whole visual system of eye and brain, then a better comparison is with the closed circuit

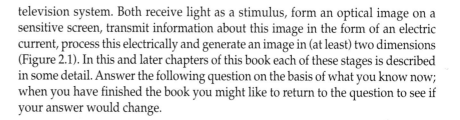

television system. Both receive light as a stimulus, form an optical image on a sensitive screen, transmit information about this image in the form of an electric current, process this electrically and generate an image in (at least) two dimensions (Figure 2.1). In this and later chapters of this book each of these stages is described in some detail. Answer the following question on the basis of what you know now; when you have finished the book you might like to return to the question to see if your answer would change.

---

**QUESTION**

**2.1** List those features of the structure and functions of the TV system that are shared by the eye. In which is the eye superior and which the TV?

---

**FIG 2.1** Vision and television – a comparison of systems.

A summary of the structure of the eye is given in Figure 2.2. The following are detailed descriptions of the functions of the more important parts.

**The cornea** is the transparent bump on the front of the eye which is responsible for about two thirds of the **refraction** in image formation, most of it at the front interface. Variations in its shape account for most of the optical defects of vision that people have; contact lenses which sit on the front of the cornea are one way of correcting these. The cornea is supplied with oxygen from the air as blood cells are impervious to light. It is self healing if scratched, but can become opaque as a result of disease or exposure to ionising radiation; corneal transplants are a successful treatment.

**The aqueous and vitreous humours** are clear liquids which provide nutrients to the parts of the eye, without interfering with the passage of light. They also provide an excess pressure to keep the eyeball firmly in shape.

**The lens** is biconvex and divides the eye into two chambers, containing the two fluids; refraction occurs at both lens surfaces. It is made of a transparent fibrous material which is added throughout life to produce a layered structure with a varying refractive index. The cells in the centre die through oxygen starvation, so the lens gets gradually stiffer as we age. It is attached by the suspensory ligaments

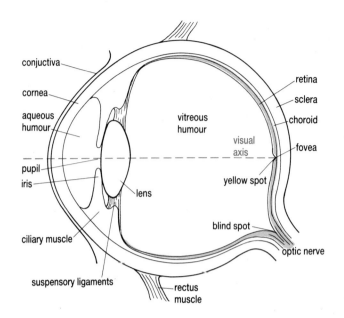

**FIG 2.2** Horizontal section of the right eye.

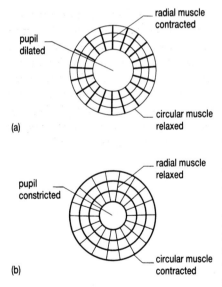

**FIG 2.3** Operation of the iris.

Trauma is the medical expression for a sudden shock, in this case physical. One of the most common causes of detached retina is to be hit in the eye with a squash ball – if you are a player, take care!

to the ciliary muscle and these together can extend or compress the lens. The refraction of the lens is changed by this so that it acts as the **fine focus control** of the eye. This automatic process is called **accommodation** and it is necessary to form clear images of objects at different distances. The lens, like the cornea, can become opaque, and develop cataracts. These can be surgically removed, and bifocal spectacles provided to cope with the loss of accommodation.

**The iris** is the eye's adjustable **brightness control**, allowing in more or less light through the aperture, the **pupil** of the eye. It consists of radial muscles which contract to dilate the pupil when illumination falls, and circular muscles which contract to constrict the pupil when the light becomes brighter (Figure 2.3). The retina can pass a message direct to the iris – a so-called feedback loop, to ensure that this response is very rapid.

**The retina** is a layered structure on the rear inner surface of the eye which is shown schematically in Figure 2.4. It consists of two types of light-sensitive cells, the **rods and cones**, so called because of the shape of their outer ends. The rods and cones are, perhaps surprisingly, situated at the back of the retina behind the jumble of connecting nerves that carry the electrical signals to the brain. There are about 120 million rods in the retina and about 6 million cones, with the cones predominating around the centre, the optical axis, and the rods being more frequent at the periphery. On the optical axis a **yellow spot** of about 1 mm in diameter can be observed with a depression in the centre called the **fovea**. This is the position of most acute vision since it contains only cones, and there are no nerve fibres obstructing the light as they are displaced radially from it. A white spot can be observed about 10° from this, towards the nose. This is where the nerve fibres collect to form the **optic nerve** to the brain. There are no receptors here so it is called the **blind spot**, and is responsible for a small field of blindness if only one eye is used. The retina can become detached from the outer layers of the eye as a result of trauma or disease. One of the successful uses of the laser is in the reattachment by local heating. This is described in Chapter 7.

**Sclera** is the external covering of the eye which is made of a tough fibrous tissue to protect the contents. Inside this is the **choroid** or pigment epithelium layer which provides blood to the retina, and is heavily pigmented to prevent reflection of light within the eye, which would cause a blurring of the image.

**FIG 2.4** Schematic structure of the retina. Several rods share a single nerve fibre, but each cone is connected to a single nerve fibre.

**Rectus muscles** hold the eyeball in the skull socket called the **orbit**. There are three pairs of muscles, one pair for each of the three directions of rotation to permit free movement. The aim is to have the **visual axis** through the fovea directed towards the object viewed, for this gives most acute vision. If the object moves towards the eyes, these muscles will automatically pivot the eyes to follow it, so the visual axes **converge**. This process can be defective, so that the images from the two eyes are not exactly superimposed, causing double vision or diplopia. This can occur especially when the eyes have to follow close movement, which may, for example, cause reading difficulties. Temporary diplopia can be caused by excess alcohol.

 ## THE RESPONSE SYSTEM

## *Photodetection*

A photon is a quantum of light energy, the smallest amount of light that can exist.

Each rod or cone acts as a photocell, releasing electric charge when light falls on it. Experiment has indicated that they can respond to a single photon. Rods and cones are responsible for different parts of our vision.

**Rods** do not register colour but are more sensitive to light than cones so can operate at lower light intensities. This is as a result of several of them converging on a single nerve fibre (Figure 2.4), which results in rods having poorer visual acuity. Rods are therefore used for **scotopic** (dim light or night time) vision.

**Cones** differentiate colour as they contain three visual pigments. They produce more acute vision because they do not share nerve fibres, but they need higher levels of light than rods. Cones are therefore used for **photopic** (bright or daylight) vision.

Eating a lot of carrots enables you to see in the dark. This was the claim made by the British during the Second World War to explain why they were able to detect German night-flying planes. This exaggeration of the truth was to conceal from the enemy the successful development of radar detectors.

Rods contain the light-sensitive pigment, **rhodopsin** (often called **visual purple** because of its appearance). This is made of a protein and a derivative of vitamin A called **retinene**, which is changed into a different form by light, in a process called **bleaching**. This reaction results in the generation of an electrical potential, and this passes down the nerve to the brain. The bleached rhodopsin is rapidly changed back into its light-sensitive form by vitamin A supplied in the bloodstream by the choroid, so that it can again respond to light. Shortage of vitamin A (retinol) makes it difficult to see in dim light. It can also cause the cornea to become thick and dry. Good sources of the vitamin are fish liver oil (ugh!) and carrots.

## Sensitivity

The maximum sensitivity of the eye has been estimated as a response to a single photon. What is also remarkable is the eye's ability to distinguish differences in intensity, over an intensity range of about $10^9$.

Full sunshine is about $10^6$ times more intense than full moonlight.

You will notice this if you try to take photographs using a camera which needs you to estimate the exposure, you will inevitably underestimate the exposure time needed in dim light and vice versa. The eye has this automatic control by a combination of the iris (total illumination), and dark adaptation.

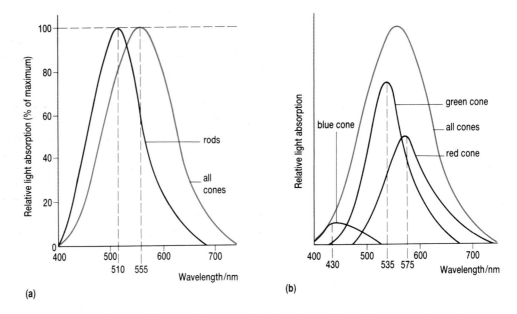

**FIG 2.5** Receptor absorption **(a)** spectral absorption of rods and cones **(b)** contributions to total cone absorption.

**Dark adaptation** is the mechanism by which the eye increases its sensitivity by up to $10^6$ as intensity falls. There are two parts to this process, the photochemical and the neural. Light on the retina causes the cells to be depleted of their photosensitive pigments in the process of response. In the absence of light, the concentration of the photosensitive chemicals in the rods and cones gradually builds up over a period of about 30 minutes, which increases the sensitivity to any dim light which is received. This occurs first for the cones and then for the rods. The second effect is the collection by the rods of energy from a greater area, by more sharing of nerve fibres. This means that smaller intensities produce a signal, at the cost of the ability to distinguish fine detail (a loss of visual acuity – see below). There is also a collection of light over a longer time. Thus the playing of ball games in dim light is difficult because of loss of accurate position and time information, though the ball can still be seen.

# Spectral response

The minimum or threshold intensity needed to see a flash of light is very wavelength dependent. The cornea is opaque to wavelengths shorter than 300 nm, and the lens to wavelengths below 380 nm. Rods and cones do not detect wavelengths above about 700 nm.

Although the rods do not give any colour information, they are most sensitive to light in the green part of the spectrum, with wavelengths of about 510 nm – see Figure 2.5(a).

Experimental evidence has now confirmed the **trichromatic** theory of colour vision. This is that there are three types of cones containing three different pigments. These are:

■ a 'red' cone containing the red-sensitive pigment **erythrolabe**,

■ a 'green' cone containing the green-sensitive pigment **chlorolabe**,

■ a 'blue' cone containing the blue-sensitive pigment **cyanolabe**.

In responding to light, each is believed to undergo a bleaching process similar to that which occurs in rods. As Figure 2.5(b) shows, the green cones are most sensitive and the blue least, giving an overall maximum sensitivity in the yellow wavelength range.

The perception of colour is a result of the response of particular cones (Figure 2.5(b)). Monochromatic light of wavelength 430 nm will stimulate only the blue cones and so be seen as blue, whereas light of wavelength 550 nm will stimulate green and red cones and so appear yellow. Light with a particular wavelength will be detected by up to three types of cone in different proportions to give the specific spectral colours.

**Colour blindness**, or more usually, defects in colour vision are caused by the shortage of a particular kind of cone. The failure to distinguish between red and green is one of the more common ones, especially in males. Certain defects are not satisfactorily explained by the trichromatic theory, nor is the occurrence of negative afterimages. This is the perception of a complementary colour following the removal of a colour after viewing it for some time. There are clearly complexities in the perception of colour that will not be fully understood until we have a much greater knowledge of the brain.

Use the information in this section to answer the following questions.

## QUESTIONS

**2.2** Why do you get sharper vision at night by looking out of the side of your eye?

**2.3** Explain light adaptation, the opposite effect to dark adaptation. How does the time taken compare with dark adaptation?

**2.4** Explain why an object emitting equal intensities of light of wavelengths 535 nm and 575 nm appears yellow.

**2.5** If you view a well-illuminated red object against a white background for about one minute, you may see a cyan (blue-green) afterimage when it is removed. Can you think of a possible explanation of this? How could you test the explanation?

**2.6** Explain why yellow is a good colour to use for tennis balls or pedestrians' warning clothing.

## Resolution

The ability of the eye to discern detail is a very important factor in vision. **Resolving power** is defined as the minimum separation angle $\theta$ (in radians) of two objects that can be separately distinguished. It can be shown that

$$\theta \approx \sin \theta = \lambda/D,$$

where $D$ is the diameter of the eye lens aperture.

**Visual acuity** is defined as the inverse of the angle $\theta$ (in minutes) by which the two nearest points are separated, that is

visual acuity $= 1/\theta$ per minute

This separation will depend on the structure of the retina. If a distinction between two points of light is to be made, then there must be at least one unstimulated receptor cell between those stimulated. Hence the minimum spacing will be twice the diameter of the receptor cell. The cells are most closely packed at the fovea and the cones have their own nerve fibre. Their diameter is about 1.5 mm, giving a minimum spacing of 3 μm, a visual angle of half a minute, giving a visual acuity of 2 per minute (Figure 2.6). This is in fact similar to the limit of resolution which is set by the diffraction of light as it passes into the eye (see the next section for details).

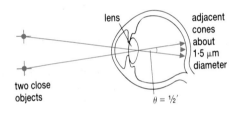

**FIG 2.6** Visual acuity.

Elsewhere in the retina the cones are less frequent and the rods share nerve fibres, so a single signal is produced by an increased area of receptors. At the periphery of the retina about 600 rods are connected to one nerve, giving an acuity of about 0.1 per minute.

$$\text{Angle in radians} = \frac{\text{arc length}}{\text{radius}}$$

## Response to illumination changes over time

When viewing an object under steady illumination, producing a fixed image on the retina, the eye executes a continuous **scanning** movement. This ensures that there are always some new nerves being stimulated to produce a signal, and avoids the fading of the image as the receptors cease to produce the chemical change necessary to generate the electrical potential.

By contrast, if the illumination changes rapidly, the excitation of the receptors takes a finite time to respond. This delay is called **persistence of vision** because you will see the light for longer than it was actually reaching the eye. The delay period depends on the intensity of the light. Bright flashes will 'flicker fuse' into a continuous brightness at a frequency of about 50 Hz. In dim light this will occur at a flicker rate of only 5 Hz, because the rods respond less quickly to the changes than the cones. This ability (or disability), of the eye is the basis of movies, in which pictures are flashed on the screen at 24 frames a second, so that the gaps in the action are filled in by the eye to produce a smooth motion. Television screens use a similar principle in the scanning of the screen by the electron beam. This is described in Chapter 4.

## SOCIAL IMPLICATIONS OF COLOUR
**2.3** PERCEPTION

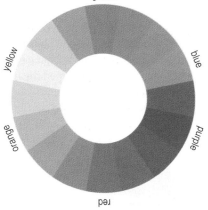

**FIG 2.7** Disc of the hues of the spectrum, and brightness.

**W**e spend our lives surrounded by light and colour; not surprisingly, this has considerable effects on us in a range of different contexts. To get to grips with this, many people have asked, how can colour be classified?

### The dimensions of colour

The spectral wavelength of light is only one attribute of colour, called the **hue**. Two others are needed to define any colour. These are **brightness**, which depends on the intensity of the light and **saturation**. This is a measure of the purity of a colour, that is how much of the intensity is from the hue, or particular wavelength, and how much from a broad spectrum of a mixing colour such as white or grey. Thus a red hue may be reduced in saturation by mixing it with white to produce pink. In order to represent the complete range of colour therefore, three dimensions are needed. The hue can be shown as a disc of the spectrum, (Figure 2.7). The radius of the sphere represents the brightness, with intensity increasing outwards. The saturation is the third dimension ranging from black to white. Putting all three together results in a double cone which is shown in a cut-away form in Figure 2.8. In the centre slice the colours are most saturated and at the outer edge of this they are brightest.

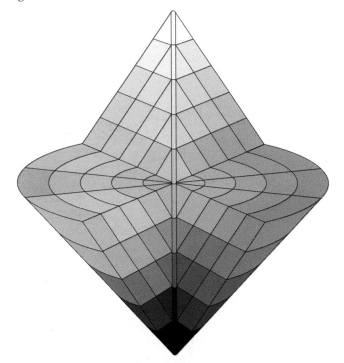

**FIG 2.8** Colour cone showing hues, brightness and saturation of colour.

The actual perception of colour is also affected by the material nature of the object viewed and its surroundings (Figure 2.9). Figures 2.7–2.13 are displayed in full colour on the inside front and back cover.

**FIG 2.9** How the surroundings affect perception – all three figures have a centre of the same colour.

## Perception effects

There is evidence that colour can affect the hypothalamus which is the part of the brain which is the centre of emotions. It affects the pituitary gland which controls hormone levels and therefore can determine moods.

Much research has been carried out to try to establish the psychological effects of colour, as opposed to the physiological effects of the hormones. Tests have been scientifically inconclusive, probably because the complexity of the factors involved mean that experimental repeatability is very difficult. For example, nursery age children have been shown to prefer yellow crayons to show happiness and brown ones when drawing sad people. Whether this is a direct effect in the brain or is socially conditioned is not clear, however.

Nevertheless it is commonly accepted that colour influences what psychologists call **affect**, the emotional side of life, rather than the intellect. This can be used to communicate feelings effectively without words. Figure 2.10 is taken from a book designed for adults who cannot read. Some psychologists have developed theories or classifications of these effects, (see box). Those who design the colour use in various aspects of everyday life may claim to be working pragmatically, but their results do conform to general patterns.

**FIG 2.10** Using colour in a story to communicate emotion or 'affect'.

## *Colour in advertising*

Imagine a visit to a supermarket. The vast array of goods that confront you need to establish their presence in a very short time – research suggests your gaze may linger on an item for only 0.03 second. Colours at the long wavelength end of the spectrum – red and yellow – are 'hot' and look closer or larger than those in blue or purple. These colours are favoured for the quick messages, such as 'special offers', 'new, improved recipe' and 'value for money' claims. Children and those with unsophisticated tastes prefer bright, primary colours, so toys and inexpensive items will display these. By contrast goods appealing to those with more sophisticated tastes tend to have cooler colours of blues and greens. Expensive items appealing to the taste for luxury are often packaged in gold and silver, black and white.

### INVESTIGATION

The Swiss psychologist Max Luscher evolved a test of colour preferences which he used to categorise psychological preferences. Four colours are called the 'psychological primaries', red, yellow, blue–green and dark blue. Products packaged in these colours, appeal to a particular need. Red offers vitality and power, and appeals to younger people, as illustrated in the drinks can (Figure 2.11(a)). Yellow offers modernity, the future; it is often combined with red to offer the power to control the future whether by photographing it or driving to it!

(Figure 2.11(b)). Blue–green appeals to a need for clarity and certainty, which is why this book uses it as a second colour! Dark blue is associated with contentment and designers have used this to offer security with products such as the first mass-produced car and financial institutions (Figure 2.11(c)).

1. Do a quick survey of a range of products, for example as advertised in popular magazines. Are these associations apparent? If not are there any others which might be variants of these?

(a)

(b)

(c)

**FIG 2.11** Colour in advertising **(a)** red **(b)** yellow **(c)** dark-blue.

## Colour in the environment

The natural environment is likely to condition us to associate certain features with particular colours, and link these features to emotions. Red, orange and yellow are the colours of the Sun, fire and blood, all of which suggest danger and power. Warning safety signs should be more visible if they are shown in these colours, but they may have an added psychological impact. The use of these colours as warnings in insects, for example yellow wasps and red ladybirds, suggests that the impact does not need a thinking brain – wasps are not just warning us, but birds that might eat them.

British troops were once famous as 'redcoats' because of their flamboyant uniforms, designed to proclaim power to the allies and danger to the enemy. In modern warfare the power is delivered by long-range weapons not the physical attributes of the soldiers. So the uniform serves to camouflage the combatant in the environment. Animals also use this strategy in the battle for survival that drives evolution. The uniforms of police and airline staff usually feature dark blue which is an example of the association of this colour with security.

Green and blue in the environment are associated with things which do not change quickly or unpredictably (though the sky in Britain is perhaps an exception to this!). They are therefore linked to passivity and tranquillity. Hospital walls are often painted in pastel shades of green (blue is rather cold), blue jeans are something that people wear to be inconspicuous in a group (or is it the power of advertising which ensures everyone wears them?).

There is a physical process which contributes to the perception of 'hot' colours appearing in the foreground and 'cool' ones in the background. Scattering of light by particles in the atmosphere is strongly dependent on wavelength. The sky is blue because of the scattering of blue light from the Sun in all directions. The Sun is red at sunset because only the red light is not scattered through the greater thickness of atmosphere it is seen through when near the horizon.

Architects and interior designers use these effects to enhance the impact of buildings and room spaces. Figure 2.12 shows two contrasting examples. In the design of the Gothic cathedral the aim was to achieve maximum height so ceilings and lofty stained glass windows would be blue so they appear to recede. In Islamic architecture, the use of patterns and shapes reduces the appearance of depth and focuses attention on the complexities of the inner space.

Colour can affect the perception of time also. For example there is research evidence that meetings seem to be shorter in brighter rooms than those in more subtle colours – perhaps you could check this out with classroom experiences!

### QUESTION

**2.7**    (a) Explain how the surrounding of the object in Figure 2.9 affects its appearance.

       (b) Figure 2.13 shows two eating places. Describe how the use of colour creates expectations of what kind of food and dining experience is on offer.

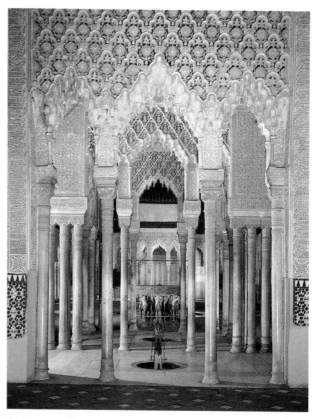

**FIG 2.12** Colour in architecture **(a)** Gothic church

**(b)** Mosque.

**FIG 2.13** Colour for ambience **(a)** fast food café

**(b)** sophisticated restaurant.

**2.4**  THE OPTICAL SYSTEM

**T**he ability of the eye to form images is as a result of refraction at the curved interfaces between the various media that the light passes through on its way to the retina. The curvature can be described as **refracting power**.

## *Refracting power*

The refractive power of a surface is defined as:

power (in dioptres, D) = 1/focal length (in metres),
$$P = 1/f \tag{1}$$

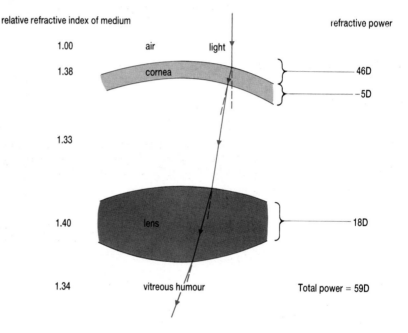

**FIG 2.14** Refraction in the eye.

Powers of a series of surfaces can be added algebraically, which makes the unit more useful than focal length. The amount of refraction is dependent on the change in the relative refractive index at the interface, and as Figure 2.14 shows this is greatest at the front of cornea. There is much less refraction (and it is in the opposite sense) at the rear, because the interface is with the aqueous humour, which has a similar refractive index. The lens produces less refraction than the cornea, but because its shape is variable it acts as the adjustable focus. When the lens is at its flattest it has a power of about 18 D. This is for producing an image on the retina of distant objects. This is called the **far point**, and for the normal eye it is at infinity. To focus on closer objects the lens needs to increase its power, to converge the beam. In young people the maximum power is about 29 D, to produce a **near point**, the nearest position of clear vision, of less than 0.25 m. Measure your own near point. The range of focus available is the eye's **accommodation** (Figure 2.15). The power of accommodation can be calculated from the difference in power, in this case,

$$P_n - P_f = 29 - 18 = 11 \text{ D}.$$

In young children the power of accommodation may be as high as 14 D, in elderly people it can fall to zero.

## QUESTIONS

**2.8**   Why is it impossible to focus your eyes under water? (*Hint* – refer to Figure 2.14).

**2.9**   Why do people lose the power of accommodation during their lives?

**FIG 2.15** Accommodation – range of focus **(a)** distance object **(b)** near object.

## Depth of field

The distance from lens to retina is the image distance and is fixed by the dimensions of the eye. For any particular lens focal length, objects at only one distance will form a sharp image. In practice there is a range of object distances which can produce an acceptably clear image. This range of distance is called the **depth of field**. It is much greater for distant objects than for near objects because the difference in divergence of the light is much greater for nearer objects (Figure 2.16). The lens has to be more powerful to produce the image. You will notice this if you use a camera with a powerful lens or look through a telescope. The shorter the focal length, the smaller is the depth of field.

**FIG 2.16** Depth of field.

**FIG 2.17** Depth of focus **(a)** small aperture **(b)** large aperture.

## Depth of focus

The divergence of rays from an object can be reduced by reducing the size of the pupil (Figure 2.17). This increases the range of image distances that can be accepted by the retina as a focused image. This is called the **depth of focus**. You will notice this using a camera if you change the f-number or aperture size. There is a depth of focus range shown for each f-number which increases as the f-number increases and the aperture reduces.

### INVESTIGATION

#### Monocular and binocular vision

Here are some simple activities to examine the differences in the images from your two eyes.

1. Look at stationary objects whilst moving your head from side-to-side. How does this effect, called parallax, help in the judgement of distance?

2. Close one eye and explore your immediate surroundings, by reaching to grasp objects or by moving amongst them. Do you have a preferred eye? Do you find any difficulty in this task?

3. Repeat this with both eyes open. Can you explain why there was less difference between the two experiences than might have been expected?

4. With one eye open, try bringing together two fingers from different hands, or two pencils. How does this compare with your previous actions, why?

5. Roll a piece of paper into a tube and look with one eye through it at a plain, well-illuminated wall. Hold your other hand alongside and look at it with your other eye. Why can you see through your hand?!

## Depth perception

Perception is a complex process using not only the image on the retina, but also the brain's memory to interpret a scene. Familiarity with the scene or object viewed, and with the effects of perspective can help avoid the limitations of monocular sight. You will notice these more in new situations where there are fewer clues for the brain.

In binocular vision, each eye has a slightly different view, because of its position. The brain is able to compare the two images to perceive the third dimension, depth. This stereoscopic effect is used in photography, and can be a useful diagnostic technique in internal medical examinations, for example using X-rays. In each of these examples, pictures are taken from two adjacent positions as two eyes would view the scene. The pictures are then placed in a mount so that each eye sees one picture, and the 3-D construction is carried out by the brain. It is also used routinely in the scanning electron microscope to give a fuller picture of what is being examined (Figure 2.18).

**FIG 2.18** Stereoscopic scanning electron images of a cat flea. (To view, look beyond the picture to relax the eyes, then focus on the right-hand mouth part of both pictures.)

As the distance from the eye to the object increases, the visual angle between objects at different distances will decrease. The smallest that can be detected is, as was discussed above, about a half minute. This means that stereoscopic vision vanishes at about 60 m, and depth perception beyond this is due to the brain's interpretation, for which one eye would be satisfactory.

Other advantages of binocular vision are the increased field of view – over 180° horizontally, avoidance of a blind spot and a reduction in the impairment of vision if one eye is damaged.

# Optical acuity

In addition to the perceptual limitations of the retina discussed above, the sharpness of the image is limited by optical factors.

**Chromatic aberration** is the blurring of the image due to light of different wavelengths (colour) being brought to a focus in slightly different positions because the refractive index depends on wavelength.

**Spherical aberration** is the result of a wide beam of light entering the eye and being brought to a focus in different positions, not all of which will be on the retina. This will clearly be more serious when the pupil is larger.

**Diffraction** occurs as the light passes through the pupil, being more pronounced as the aperture becomes smaller.

The result is that the image of a point source of light is a spot with a maximum intensity at the centre and a blurred edge. The optimum size for the pupil is about 5 mm diameter, a compromise between the competing requirements of aberration and diffraction. This results in a spot of about 10 μm diameter. The images of two object points may then overlap.

Rayleigh's criterion states that these two images are just resolved as separate points if the maximum of one intensity curve coincides with the first minimum of the other (Figure 2.19(a)).

The width of the spot, and therefore the resolution will depend on the wavelength of the light. It can be shown that the resolving power $\theta = 1.22\lambda/D$ where $D$ is the diameter of a circular aperture. This occurs under optimum conditions when the centres of the two spots are 2 μm apart on the retina. This corresponds to the best resolution possible from the arrangement of the receptors in the retina (the cones in the fovea) so the two processes are perfectly matched.

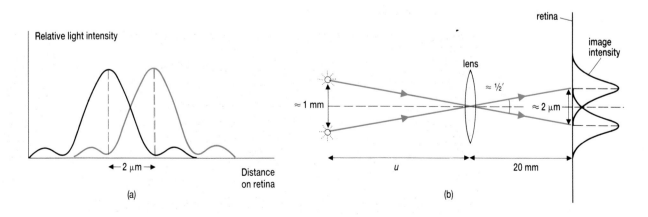

**FIG 2.19** Diffraction in the eye **(a)** the Rayleigh criterion **(b)** the eye's resolution.

**2.10** By considering similar triangles in Figure 2.19(b) find the maximum distance that this analysis predicts the eye could distinguish two points placed 1 mm apart. How does this compare with what you can actually see?

# OPTICAL DEFECTS AND THEIR CORRECTION

## 2.5

There are four common types of defect in image formation by the eye, which can be corrected by spectacles or contact lenses. These are the failure to produce a sharp image of a distant object (myopia), or a near object (hypermetropia), failure to accommodate (presbyopia), and failure to focus in certain planes (astigmatism).

## *Myopia (short sight)*

Near objects are clearly seen but the image of a distant object is formed in front of the retina, causing blurring of distant vision. The far point has moved from infinity to somewhere quite close to the eye so the range of vision has moved in towards the eye. It is a problem which often arises at adolescence, when the body grows quickly, so the eyeball is too large for the lens. In effect the power of the eye is too great and the problem is corrected with a diverging lens with a negative power (Figure 2.20(a)). This is calculated by finding the power of the eye at its far point and then selecting a lens with a power to move the far point to infinity. The thin lens formula

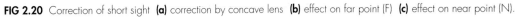

**FIG 2.20** Correction of short sight **(a)** correction by concave lens **(b)** effect on far point (F) **(c)** effect on near point (N).

$$\frac{1}{f} = \frac{1}{u} + \frac{1}{v}$$

is used, where $u$ is the object distance and $v$ the image distance, respectively, from the lens. This is combined with the expression for power $P$, (equation 1 in section 2.4) hence

$$P = \frac{1}{f} = \frac{1}{u} + \frac{1}{d} \tag{2}$$

where $d$ is the distance from the eye lens to the retina, ~0.02 m.

**(a)** Consider a man whose far point is 0.5 m. The power of his eye when fully relaxed is:

$$P_f = \frac{1}{0.5} + \frac{1}{0.02} = 52 \text{ dioptres}$$

To have his far point at infinity, the man needs the power to be:

$$P_f = \frac{1}{\infty} + \frac{1}{0.02} = 50 \text{ dioptres}$$

When wearing glasses, the effective power is the algebraic sum of the powers of his eyes and his lenses, assuming the lenses are close to the eyes, so the power of the lenses needs to be

$50 - 52 = -2 \text{ dioptres}$,

so that when his eyes are relaxed he will have a net power of 50 dioptres, and see distant objects clearly (Figure 2.20(b)).

(b) A short-sighted person will typically have the near point closer than for the normal eye. Suppose the man has an accommodation power of 5 dioptres.

$$P_n - P_f = 5$$

So his power at the near point $P_n$ is $52 + 5 = 57$ dioptres, and he will focus at a point N, Figure 2.20(c) given by:

$$P_n = \frac{1}{n} + \frac{1}{d} \quad \text{so} \quad 57 = \frac{1}{n} + \frac{1}{0.2 \text{ m}} = \frac{1}{n} + 50 \text{ dioptres}$$

Therefore the near point is at 0.14 m from the eye.

(c) Wearing the lenses will move the near point further away to N'. The power is now that of the eye at maximum power combined with the lens,

$$P_{n'} = P_n + (-2) = 57 - 2 = 55 \text{ dioptres}$$

so the new near point is given by:

$$P_{n'} = \frac{1}{n'} + \frac{1}{0.02} \quad \text{that is, } 55 = \frac{1}{n'} + 50 \text{ dioptres}$$

Which gives the new near point 0.2 m from the eye. The man may wish to remove his spectacles for close viewing.

---

### QUESTION

**2.11** (a) What is the focal length of the corrective lens used in the above example?

(b) (i) What power of lens is required to correct the short sight of a woman whose far point is 0.2 m?

(ii) What is her near point without the spectacles?

(iii) What is her near point with the spectacles?

---

## Hypermetropia (long sight)

This is the opposite of the previous defect. Distant objects can be seen clearly but the eye is insufficiently powerful to produce an image of nearer objects on the retina (Figure 2.21(a)). The range of vision has moved out from the eye. It is a common problem in infancy when the eyeball is too small for the optical system. It is corrected by an additional converging lens, with a power which will bring the near point to that of the normal eye, 0.25 m (Figure 2.21(b)).

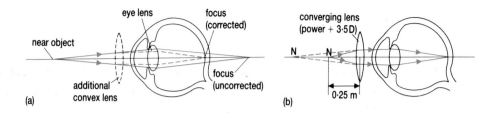

**FIG 2.21** Correction of long sight **(a)** correction by convex lens **(b)** effect on near point (N).

A radical approach to optical defects was pioneered in the USSR (Russia) and is becoming available in Western countries. This is to change the shape of the retina surgically, using the laser. Other examples of laser surgery are given in Chapter 7.

Consider a woman whose near point is 2 m from her eyes. Using equation 2 the power of her eyes is:

$$P_{n'} = \frac{1}{2} + \frac{1}{0.02} = 50.5 \text{ dioptres}$$

To focus at 0.25 m she will need a power

$$P_{n'} = \frac{1}{0.25} + \frac{1}{0.02} = 54 \text{ dioptres}$$

so the lens must provide a power of +3.5 dioptres, to give her a near point of 0.25 m.

## Presbyopia ('old sight')

With age, the lens loses its flexibility and power of accommodation. There is also a weakening of the ciliary muscles that control the lens. Older people may require a lens for both distant and close viewing. This can be provided as bifocals in which the upper part of the lens is used for distant vision and the lower part for near vision. For example an elderly myopic person may require a strongly diverging lens in the upper part and a weaker one in the lower part.

## Astigmatism

**FIG 2.22** A chart for testing astigmatism.

This is usually caused by the cornea not being spherically curved, so that it has different curvatures in different directions. Images are seen in sharper focus in one direction, e.g. vertical, than in others. It is corrected by lenses that have a cylindrical curvature in the correct orientation to compensate for the cornea. This is often combined with a curvature to correct short or long sight.

## Contact lenses

These are lenses placed on the surface of the eye to perform the same function as spectacle lenses. There are two main kinds, hard and soft. Hard lenses are made of rigid plastic, about 1 mm thick and 1 cm in diameter. They are separated from the cornea by a layer of tears. This enables the lens to take up an orientation which can compensate for astigmatism. Soft lenses are rather larger and more comfortable to wear. They conform to the shape of the cornea so they do not correct for astigmatism. They are also more easily damaged. A newer, more expensive soft lens is the toric type which can maintain its shape and so correct for astigmatism, though at the cost of some loss of comfort.

**2.12** Tabulate the differences between rods and cones in terms of their structures, functions and contributions to vision.

**2.13** Diagnose the following defects of vision and suggest, where possible, suitable corrections, including the power of the lens and any advantage of a contact or spectacle type:
   (a) Far point at infinity, near point at l m, power of accommodation 10 D.
   (b) Far point at 5 m, near point at 0.5 m, accommodation 1.8 D.
   (c) When viewing the chart shown in Figure 2.22, the vertical and near vertical lines are clearer than the horizontal lines.
   (d) Normal near and far points with each eye, double vision for objects closer than 1 m.

**2.14** Explain the following observations (you might try them out first!)
   (a) When the red filament of a lamp is viewed through a blue filter, two images, red and blue, are produced side by side.
   (b) Two objects about 10 cm apart are viewed by the right eye looking directly at the left object. When the eye is about 0.5 m away, the right hand object disappears from view.
   (c) A printed page can be read clearly when it is placed nearer than the near point, if it is viewed through a small hole. This clarity is lost if the hole has a diameter of less than about 1 mm.

**2.15** The perception of intensity of light and of colour has social implications in advertising.
   (a) Suggest a *colour* and a description of the *intensity* of light for each of the following scenes:
      (i) a hot chocolate advertisement indicating a warm and cosy atmosphere,
      (ii) a snow scene indicating a cold and exciting atmosphere.
   (b) State the response of a person to sudden changes in intensity of light, and give an everyday example of the use of such changes.

*(UCLES Feb 1997)*

**2.16 (a) (i)** Draw a **labelled** ray diagram to show how a normal eye forms an image of a near point object on the optical axis of the eye. The diagram should show the path of two rays from the object and should indicate clearly the places where refraction occurs.
      **(ii)** Explain why most of the refraction occurs at one of these places.
   **(b)** Draw another labelled ray diagram to show where the image of the same object would be formed by an eye suffering from the defect known as hypermetropia or 'long-sight'. Indicate on your diagram how a suitable correcting lens could be used to bring the image in focus on the retina.
   **(c)** A person suffering from hypermetropia has a near point at a distance of 1.25 m from her eyes.
      **(i)** Calculate the power of the correcting lens needed to bring her near point to 0.25 m from her eyes.
      **(ii)** State whether this power is positive or negative.

*(AQA Specimen 2001/2)*

**2.17** The simplified ray diagram (Figure 2.23) shows how a normal eye forms, on the retina, a point image, I, of a point object, O. The image is sharp and well focused.

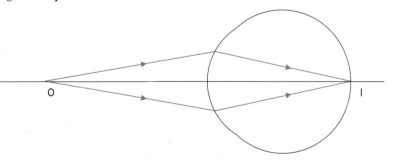

**FIG 2.23**

   (a) Explain what is meant by the *depth of field* of the eye.
   (b) Copy Figure 2.23 and mark on the diagram the positions of **two** additional point objects, $O_1$ and $O_2$, and the depth of field of the eye which they define.

To do this you will need to draw rays from the additional point objects, $O_1$ and $O_2$, showing where the eye forms their corresponding images $I_1$ and $I_2$.

**(c)** What effect does the aperture of the eye have on the depth of field of the eye and how does this affect vision at night?

*(AQA Mar 1999)*

**2.18** A student closes one eye and focuses the other on an object at infinity. The power of the open eye when focused at infinity is 59 D. The student's eye then accommodates fully to allow text to be read clearly at a distance of no nearer than 60 cm.

**(a)** **(i)** Explain the meaning of the term *accommodate*.

**(ii)** State how accommodation is achieved in the human eye.

**(b)** **(i)** Calculate the effective focal length of the student's eye when viewing an object at infinity.

**(ii)** Calculate the power of the student's eye when fully accommodated.

**(c)** **(i)** State the eye defect from which this student suffers.

**(ii)** State the type of spectacle lens which corrects this defect.

*(OCR June 1999)*

# 3

# THE EAR AND HEARING

**W**hich sense is more valuable to you, seeing or hearing? Many people who are totally deaf claim that they would rather be blind. What they have lost is our basic means of communication; in a silent world it is difficult to learn to speak and language itself is at risk.

The ear is a remarkable organ. It has a sensitivity to sound vibrations with a loudness range of more than $10^{12}$ (1000 times greater than the range of light intensities that the eye can see). It has a frequency range which varies by a factor of $10^3$, compared to the eye's range of a factor 2. For a mechanical analogy this might be compared with a weighing machine which could weigh just as accurately a flea or an elephant, whether they were jumping on and off the scales 20 or 20 000 times a second! Amongst its other impressive functions are the selectivity, which enables a person to hear a quiet word in a noisy disco; the recording of the complexity of sound so that the brain can interpret a false note in a musical performance; and the instant feedback that enables a person to monitor their own speech. In this chapter you will find out how the ear works, and what can impair its function.

## LEARNING OBJECTIVES

*After studying this chapter you should be able to:*

① give an outline description of the structure of the ear and how it transmits and detects vibration;

② describe how the ear responds to the characteristic properties of sound waves, including frequency, intensity and loudness;

③ describe the common defects of hearing and their possible remedies;

④ explain the deterioration of hearing as a result of age and exposure to excessive noise.

**T**he ear is designed to convert weak mechanical vibrations of air into electrical pulses that can be sent to the brain. It consists of a mechanical collection and amplification system, (in the outer and middle ear) and transducers to produce electrical potentials in nerves (in the inner ear). The auditory nerves lead to the auditory cortex, the part of the brain which interprets the signals. We will now follow the sequence of events involved in hearing, in the structures shown in Figure 3.1.

## The outer ear

The visible part of the ear, the **auricle** or **pinna**, is not strictly part of the outer ear as it plays very little part in the process. In some animals it has a role in collecting the sound vibrations, but in humans it can be removed without noticeable loss of hearing. This is because of the small size. Hearing can be improved by cupping your hands behind your ears, or by use of an ear trumpet, to make up for this deficiency.

The outer ear consists of the **external auditory canal** or **meatus**, which leads from the auricle to the ear drum or **tympanic membrane**. It is about 2.4 cm long and about 7 mm in diameter. It acts as an organ pipe closed at one end, so that the air in it can vibrate. This vibration is passed to the tympanic membrane, which is paper thin (~0.1 mm), with an area of about 65 mm$^2$. This behaves, as its name indicates, like a drum, though the size of the vibrations can be as small as $10^{-11}$ m. Although its fibrous structure is relatively strong, it can be ruptured by loud sounds – or pencil poking!

One of the first medical physicists to study the ear and hearing was Hermann von Helmholtz (1821–1894). His theory was developed by Georg Von Bekesey (1900–1970), a communications engineer who received the Nobel prize in 1961 for his contribution. There is still much that is unknown, particularly about the later stages of the process of hearing.

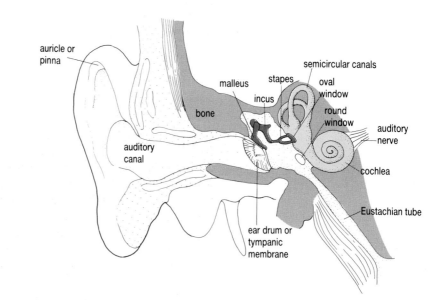

**FIG 3.1** Vertical section through the ear.

## The middle ear

There is a mechanical linkage of three small bones, the **ossicles**, between the ear drum and a smaller membrane, called the **oval window**. The bones are called the

**malleus**, **incus** and **stapes**, or, reflecting their shapes, the hammer, anvil and stirrup. They act as combined lever and pistons in the air-filled cavity between the membranes. The lever has a mechanical advantage of about 1.5 and the ratio of the areas of the ear drum and the oval window is about 15 to 1. This results in the pressure applied to the oval window being increased by a factor of more than 20, (Figure 3.2).

When sound is transmitted into a different medium a proportion of the sound is reflected, depending on a property of the medium called the **impedance**. If the impedance is similar, or matched then this avoids the loss of energy by reflection. The middle ear is designed to produce good matching.

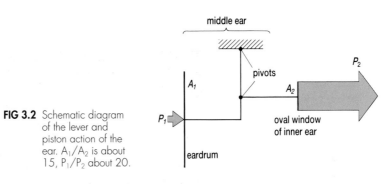

**FIG 3.2** Schematic diagram of the lever and piston action of the ear. $A_1/A_2$ is about 15, $P_1/P_2$ about 20.

The ossicles also help to protect the ear from damage by loud sounds, by a movement sideways which is not transmitted to the inner ear. This is too slow a reaction for a sudden increase in intensity. Another type of protection is given by the **Eustachian tube**, which connects the middle ear to the mouth. This enables the pressures on each side of the ear drum to be kept equal. Differences in pressure impair the hearing and can cause pain. The tube is usually closed and is opened by swallowing, yawning or chewing. You will have experienced the relief this brings when you are subjected to a sudden change in air pressure, perhaps in a plane, lift or car in hilly country.

## The inner ear

This is where things get rather complicated!

The inner ear is a cavity deep within the skull which contains two organs. The **semicircular canals** are narrow, fluid-filled tubes which are not for hearing but for the sense of balance and body movement. The **cochlea** is also a fluid-filled tube of three chambers, coiled and about the size of the little finger nail (Figure 3.3(a)). If it were uncoiled the tube would be about 3 cm long (Figure 3.3(b)). The pressure wave from the oval window passes through fluid called **perilymph**, down the spiral in the **vestibular chamber** to the end and returns via the **tympanic chamber**. The pressure variation is absorbed at the return end by the round window membrane. Between these two chambers is the **cochlear duct** or **median chamber**, filled with **endolymph** fluid. This contains the sensors in which pressure waves generate electrical signals.

This is a complex process which is not yet fully understood, involving the hair-like structure of the **organ of Corti** (Figure 3.3(c)) which is attached to two membranes running the length of the spiral. The **basilar membrane** consists of fibres which are short (~0.04 mm) and tense at the base near the oval window, increasing in size (up to 0.5 mm) and becoming slacker at the narrow end. As the pressure wave travels along the chambers from the oval to the round window it sets the basilar membrane into vibration, if the frequency is greater than 20 Hz. This movement is transmitted to the hair cells which brush against the **tectorial membrane**. The liquids on either side of these membranes, the **perilymph** and the

**endolymph** have different ionic compositions which results in a potential difference across the membrane of about 150 mV. The shearing action of the hairs and their interaction with this membrane changes this electrical potential and causes a current to flow in the nerve fibres at the base of the hair cells, which is transmitted as information to the brain. A more detailed account of the generation of potentials in nerve fibres is given in Chapter 5, if you wish to pursue the story further.

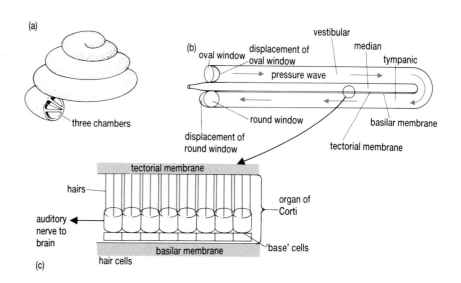

**FIG 3.3** The structure of the cochlea **(a)** external appearance **(b)** operation of 'unrolled' chambers **(c)** the organ of Corti.

---

### QUESTIONS

**3.1** Explain why you may find your hearing is impaired when you have a cold. It can also add to your discomfort to fly in an aeroplane.

**3.2** It often helps in remembering a complex and detailed account like the one just given, if you can devise a simple summary, in the form of a diagram, showing the stages in the process without all the structural details. Try this now.

---

## THE TRANSMISSION AND MEASUREMENT OF SOUND

**3.2**

Sound is a longitudinal pressure wave which is propagated by the oscillations of particles of the medium through which it is transmitted. The ear's response to sound is determined by certain characteristics of the sound, which depend on properties of the wave motion. These are:

■ pitch, which depends on the frequency,

■ loudness, which depends on intensity and on frequency,

■ quality (or timbre), which depends on the range of frequencies and their relative amplitudes.

Musical notes are usually mixtures of related frequencies such as a fundamental and harmonics. Noise is a random mixture of many unrelated frequencies.

## *Transmission of sound in a medium*

The velocity $c$ of a sound wave in a medium depends on the density $\rho$, of the medium and its bulk modulus $E$. The relationship is given by the equation:

$$c = \sqrt{E/\rho}$$

For a gas $E = \gamma p$ where $\gamma$ is the ratio of specific heats at constant pressure and constant volume, and $p$ is the pressure, so

$$c = \sqrt{\frac{\gamma p}{\rho}}$$

There is a resistance to the passage of a sound wave through a medium which is analogous to electrical resistance. This is called the **acoustic impedance** $Z$, which is given by:

$$Z = \rho c$$

This is an important quantity in determining how much sound is transmitted from one medium to another. If the impedances of the two media are very different, for example a gas and a liquid, then much of the sound will be reflected, not transmitted at the interface. The media are said to be acoustically mismatched. The ear has to arrange for acoustic matching when the sound passes from air to liquid at the middle–inner ear interface. Matching is also very important in ultrasonics as described in Chapter 8.

The **intensity** $I$ of a wave is defined as the power $P$ per unit area of the wavefront:

$$I = P/A$$

It can be shown that the intensity is related to the amplitude $a$, and the impedance:

$$I = a^2/2Z$$

That is, the intensity is proportional to the square of the amplitude, or the maximum change of pressure.

As a wave passes through a medium it will gradually lose intensity by **attenuation**. This is the result of a number of processes of interaction between the wave and the medium, including absorption, diffraction, scattering and others. As with any wave transmission, the intensity falls by a constant fraction for each unit of distance travelled, which leads to an exponential fall of intensity with distance. Intensity at a distance $x$ from the original intensity $I_0$ is given by

$$I_x = I_0 e^{-\mu x}$$

where $\mu$ is called the attenuation coefficient. Similarly an **absorption coefficient** $k$ can be defined to describe the effect of that part of the attenuation.

---

**QUESTION**

**3.3**　Write down the expression for the intensity at depth $x$ of a wave of incident intensity $I_0$ in a medium of absorption coefficient $k$.

# Intensity levels

Atmospheric pressure is approximately $10^5$ pascals, so the ear can detect a change in atmospheric pressure of two parts in ten thousand million.

The lowest intensity of sound that the ear can detect is called the **threshold**. It is typically about $10^{-12}$ W m$^{-2}$, corresponding to a pressure change of about $2 \times 10^{-5}$ pascals. The highest intensity, beyond which the ear may be damaged is reckoned to be about 100 W m$^{-2}$. This is an immense range, and it is observed that the ear's perception of a sound of intensity $I$, the loudness $L$, depends on the relationship between this intensity and the threshold intensity $I_0$ according to the expression:

$$L \propto \log_{10}(I/I_0)$$

This is because the ear responds to differences in intensity rather than total intensity, as explained in the next section.

The **intensity level** is defined in order to quantify sounds in a way which relates to the ear's response. The threshold intensity $I_0$ is taken as $10^{-12}$ W m$^{-2}$, and the intensity level of a sound of intensity $I$ W m$^{-2}$ is given in bels (B), as

$$\text{intensity level} = \log_{10}(I/I_0) \text{ B}$$

For example, if the intensity is ten times the threshold,

$$\text{intensity level} = \log_{10}(10^{-11}/10^{-12}) \text{ B} = 1 \text{ B}$$

The bel (named after Alexander Graham Bell, not the ringing device), is therefore rather a large unit, so it is more common to use one tenth, the **decibel** (dB), as the unit, so that

$$\text{intensity level} = 10 \log_{10}(I/I_0) \text{ dB} \tag{1}$$

The sound intensity of normal conversation (speakers about 1 m apart) is about $10^{-6}$ W m$^{-2}$, so the intensity level is

$$10 \log_{10}(10^{-6}/10^{-12}) \text{ dB} = 60 \text{ dB}$$

Doubling of the intensity of a sound increases the intensity level by ~3 dB. To combine two sounds, for example an alarm clock at 1 m (level 80 dB) with conversation, it is necessary to convert to intensities:

$$\text{Conversation: } I_c = 10^{-6} \text{ W m}^{-2} \text{ or } 10^6 \times I_0$$

$$\text{Alarm clock: } I_a \text{ is given by } 80 = 10 \log_{10}(I_a/I_0)$$

$$I_a = 10^8 \times I_0$$

so the total intensity $I_t = I_c + I_a = 1.01 \times 10^8 \times I_0$

and the intensity *level* $= 10 \log_{10}(1.01 \times 10^8)$ dB

$$= (10 \log_{10} 1.01 + 80) \text{ dB}$$

that is very little more than 80 dB. The alarm clock completely drowns the conversation, as you would expect from the intensities, though this may be masked by the logarithmic nature of the decibel scale, for intensity levels.

The *difference* in intensity *levels* between two sounds of intensities $I_1$ and $I_2$ is found by subtraction:

$$10 \{\log_{10}(I_2/I_0) - (I_1/I_0)\} \text{ dB}$$

which simplifies to

$$10 \log_{10}(I_2/I_1) \text{ dB} \tag{2}$$

because the scale is based on comparisons.

Thus the difference in intensity level between a conversation and a door slamming ($I = 10^{-4}$ W m$^{-2}$) is

$$10 \log_{10}(10^{-4}/10^{-6}) \text{ dB} = 20 \text{ dB}$$

## QUESTION

**3.4** (a) The upper limit of acceptable noise, by law is $10^9$ times the threshold intensity. What is its intensity level in dB?

(b) The intensity levels that can be found in a classroom range from 30 dB (rustle of paper during exams!) to 80 dB (everyone talks at once). Express this as a ratio of intensities.

**3.3**

# THE EAR'S RESPONSE

## *Frequency response*

The physical response of the ear to sound is essentially one of resonance. That is the vibrations of the sound match the natural frequencies of vibration of parts of the ear. The outer ear is a tube of length about 2.5 cm with one end closed. Its natural frequency is for a standing wave of length 10 cm to be produced (see a core physics text for details). Since

$$c = f\lambda$$

$$f = 330 \text{ m s}^{-1}/0.1 \text{ m} = 3300 \text{ Hz}$$

The middle ear displays a broader range of resonance between about 700 Hz and 1500 Hz. Within the inner ear the pressure variation becomes a travelling wave within the fluid. The amplitude of the wave decays as it travels and this depends on the frequency of the wave. High frequencies are absorbed in a short distance, low frequencies in a longer distance. This causes different parts of the basilar membrane to be affected by particular frequencies. Thus the short fibres of the organ of Corti, near the oval window respond to high frequencies, while the longer fibres at the narrow end detect the low frequencies. Below 20 Hz, there is no stimulation of the hairs. The brain discriminates frequencies by distinguishing the places from which the impulses come. The combination of these various resonances gives the ear its **frequency range**. The high frequency limit for hearing is about 20 kHz, though this falls with age. Frequencies beyond this are called ultrasonic.

## QUESTIONS

**3.5** The data on the frequency response of the ear is summarised in Figure 3.4. Use this to write down:

(a) the intensity of the minimum threshold of hearing and the approximate range of frequencies over which the ear has this sensitivity. How does this relate to the structure of the ear?

(b) intensity levels of discomfort and pain; are these frequency dependent?

(c) the approximate audible frequency range of sound level 60 dB.

**3.6** If you have access to a sound intensity level meter and an audio signal generator, measure your own audio response.

**Frequency discrimination** is the ability to distinguish one frequency from another. The ear's response in this is also frequency dependent. It is greatest at low frequencies; in the range 60–1000 Hz a difference of about 3 Hz can be distinguished. This falls away until at above 10 kHz it is very poor. This behaviour of the ear has resulted in the development of musical intervals, which depend on the ratio of frequencies rather than their differences. An octave is a ratio of upper and lower frequencies of 2:1. For example middle C has a frequency of 256 Hz, one octave above this is upper C with a frequency of 512 Hz. It is called an octave because there are 8 notes spanning this frequency range, named A to C and an additional C. Each pair of notes are an interval in which the two frequencies are in a certain ratio.

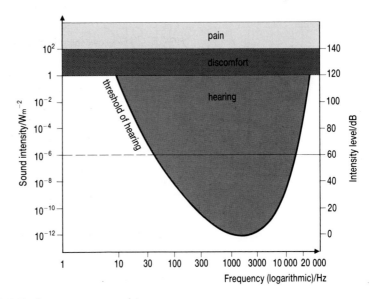

**FIG 3.4** The frequency response of the average ear.

If two notes close in frequency are sounded simultaneously, the waves will interfere with each other. This produces a regular variation in intensity which is heard as **beats**. The beat frequency is the number of intensity maxima which are heard each second. It is equal to the difference between the two individual frequencies. The ear can detect beat frequencies of 1–6 Hz, largely independent of the original frequencies.

## Intensity response – loudness

The ear's response to an increase in intensity has at least three parts. In the inner ear the greater amplitude of the wave results in

**(a)** a greater movement of the basilar membrane, producing more stimulation of the nerve endings by the hair cells;

**(b)** additional hair cells are activated to stimulate nerve endings, in the particular location for that frequency of sound;

**(c)** nerves are stimulated beyond that part of the membrane as a result of its greater movement.

Each of these effects increases the extent of the nerve impulse which is sent to the brain. This increase is perceived as an increase in loudness $L$.

Thus loudness is dependent on intensity but is also a result of the energy transfer characteristics of the ear. It is measured in **phon**, and like the threshold of hearing, is strongly frequency dependent for a given intensity (Figure 3.5).

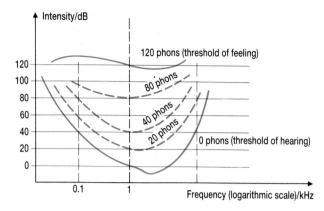

**FIG 3.5** Loudness perception by the average ear.

The perception of changes in loudness does not correspond directly to changes in intensity. Equal increases in loudness result from equal proportional changes in intensity. For example if the intensity is doubled say from $1 \times 10^{-6}$ to $2 \times 10^{-6}$, and from $2 \times 10^{-6}$ to $4 \times 10^{-6}$, the loudness will be heard to increase by the same amount each time. This can be expressed as

$$\text{loudness increase} \propto \frac{\text{intensity increase}}{\text{initial intensity}}$$

$$dL \propto \frac{dI}{I} \text{ or } dL = k\frac{dI}{I}$$

where k is a constant. Integrating this gives

$$L = k \log_e I + C$$

where C is also a constant. It may be determined as follows:

when $I = I_0$ the threshold intensity, the loudness is zero, $L = 0$, so

$$C = -k \log_e I_0$$

therefore $L = k \log_e I - k \log_e I_0$

$$= k \log_e(I/I_0)$$

or $\qquad L \propto \log_{10}(I/I_0)$

which expresses the ear's **logarithmic** response to intensity changes. This relationship is an example of the Weber–Fechner law, which states that the change in the perceived property is proportional to the *fractional* change in the stimulus. It is common to a number of stimuli including weight and brightness.

The **sensitivity** of the ear is its ability to detect the smallest fractional change in intensity $dI/I$. It is defined as

$$S = \log (I/dI)$$

Sensitivity, as you would expect, is strongly dependent on frequency and intensity, being a maximum for low intensities at about 2 kHz. The minimum change in intensity that can be detected is about 0.5 dB. This corresponds to a ratio of sound intensities $I_1/I_2$, given by equation (2) as

$$0.5 = 10 \log_{10}(I_1/I_2)$$

$$I_1/I_2 = \text{antilog}_{10}(1/20)$$

$$= 1.12$$

an increase in intensity of 12 per cent or one eighth. This is comparable with the sense of weight change at ~10 per cent but is considerably poorer than the sense of brightness change at ~1 per cent.

A variation on the 'guess the weight of a cake' competition might be to find the largest slice that can be removed without it being noticeable to someone who is blindfolded. By feeling the weight most people could tell when a tenth has been removed.

As previously described the intensity level scale is defined to be similar to the ear's response, that is logarithmic. This describes a physical quantity, however, which can be calculated from a knowledge of the amplitude or intensity of the wave. Loudness is a subjective sensation and so has to be measured subjectively. Since loudness is frequency dependent, measurement must take this into account. There are two ways in which this is done.

In **phon measurements** all sounds are referred to a standard frequency of 1 kHz. In order to measure the loudness of an unknown source the following procedure is followed. The source to be measured is placed near the standard source, whose intensity is adjusted until the two sources have equal loudness. If the standard source has an intensity level of $l$ dB, then the loudness of the unknown source is $l$ phons. This is not a very practical procedure in many situations so there is a more common method.

The **dBA scale** is an adaptation to the intensity level scale to take account of frequency dependence. Measurements are made with a sound level meter which is a microphone calibrated to register a 'weighted decibel' reading. This is done by means of circuits which suppress the contributions of high and low frequencies, so that the response is similar to that of the ear. This scale is now used almost universally in the measurement of noise, since it maintains the relationship between perceived loudness and measured value of 'weighted intensity', for example, an increase of 10 dBA doubles the loudness. Some values of typical noises are given in Table 3.1.

**TABLE 3.1** Intensity levels of some typical noises

| LEVEL/dBA | SOUND (AND EFFECT) |
|---|---|
| 160 | Rifle close to ear (ear drum ruptures) |
| 140 | Aircraft at 25 m (threshold of pain) |
| 120 | Discotheque, close to speakers (threshold of discomfort) |
| 100 | Very noisy factory, noisy food blender at 0.5 m |
| 90 | Road drill or heavy lorry at 7 m, underground train (upper limit of acceptable noise level, by law) |
| 70 | Busy street, vacuum cleaner at 3 m |
| 50 | Quiet street, 'normal' rooms at home |
| 40 | Quiet conversation |
| 30 | Whisper, tick of watch |
| 20 | Blood pulsing |
| 0 | Threshold of hearing |

The damage that sounds cause depends on the length of time the ears are exposed for as well as the intensity. Sounds or noise rarely continue for long at the same intensity, so a further quantity is defined to enable comparisons to be made of the effects of varying levels. The **equivalent continuous sound level** $L_{eq}$ is that sound level which if constant for a defined period, would give an exposure to sound energy equivalent to that received. The defined period is usually 8 hours, as that is the nominal length of the working day. The quantity is thus called $L_{eq}$[8 h]. It is measured in dBA.

## QUESTION

**3.7**  From Table 3.1 and values previously stated, calculate:
(a) The ratio of intensities between the threshold of hearing and ear drum rupture.
(b) The intensity of sound to cause ear drum rupture, in W m$^{-2}$.
(c) The change in intensity when an aircraft flies overhead at 25 m, when you are walking in the quiet countryside.

## Stereophonic hearing

The localisation of a sound is possible because of the differences in the sound received by each ear. This is more difficult with low frequency sounds because of increased diffraction effects which cause the sound to be more diffuse. The brain can make use of three differences between the signals received by each ear.

(a) There is a time lag which is dependent on position, though there can be some confusion between locations giving the same time lag. This is particularly useful for short, sharp sounds and low frequencies.

(b) There is a phase difference (the two waves are at different stages in the compression–rarefaction cycle). This is important in continuous sounds and low frequencies.

(c) There is an intensity difference, because of distance attenuation and shadow effects of the head. This is most useful for high frequencies because diffraction effects are small.

---

**QUESTION**

**3.8**  (a) Why is diffraction less common in high frequency sound? How does diffraction interfere with the location mechanisms?

(b) Speech is located using the first two mechanisms described above, why?

(c) Explain why the first method is useful for short sounds and the second for long ones.

(d) Find out which are the recommended positions for speakers for stereophonic recordings. How do these relate to the binaural hearing mechanisms?

---

## 3.4  DEFECTS OF HEARING

### Causes

Hearing losses are usually divided into two categories. **Conductive** loss in which the sound vibrations do not reach the inner ear, and **sensorineural** loss where the cochlea does not pass impulses to the brain. The main causes are accident or trauma, disease, age and exposure to excessive noise. Young children are susceptible to temporary deafness as a result of cold-type infections in which fluid fills the middle ear. The ear drum can be ruptured by a sudden shock wave, the inner ear can be damaged by a blow on the head. Disease and age can reduce the ability of the bones of the middle ear or the oval window to respond to the pressure variations. Conductive hearing loss can sometimes be corrected by surgery, such as the replacement of solidified ossicles with plastic ones. An alternative is the use of a hearing aid which transmits the vibrations through the bones of the skull to the inner ear. Electronic aids (and ear trumpets) amplify the vibrations to compensate for the loss of sensitivity of the ear. Usually the frequency range is very limited, for conversation 200–5000 Hz is adequate. This compares with the telephone's range of 30–3400 Hz.

The usual cause of sensorineural hearing loss is damaged hair cells. The auditory nerve does not therefore receive adequate electrical impulses. In extreme cases this can be treated by a cochlear implant which directly stimulates the auditory nerve to give the simulation of a sound.

There is another hearing defect whose cause is not well understood. **Tinnitus** is a constant ringing or hissing in the ears, which interferes with the reception of sound and can cause serious insomnia. It may be produced temporarily by a few hours' exposure to a noisy factory or pop concert. But if this is regularly repeated the effects may not wear off and the sufferer may not notice the steady deterioration by getting used to the internal background noise.

---

### QUESTION

**3.9** Why would a person with an effective hearing aid suffer little loss in reception of telephone sound, but be wasting their money on Hi-Fi audio equipment?

---

## Noise

Kathy Peck is a rock musician who discovered that after four years of rehearsals and performances, she had a 40% hearing loss. She set up HEAR (Hearing Awareness Education for Rockers) to warn other musicians of the hazards. Their advice includes the recommendation to anyone playing or listening to loud music, to wear earplugs.

Much is reported about the exposure of people to noise and the effects this has. This is because there is a concern that noise levels are increasing and also that the effects are not fully understood. The effects are certainly complex, ranging from the instantaneous physical damage of the ear drum or ossicles at levels of 190 dBA (close to a large gun at explosion), to the psychological effects of exposure to low frequency, low intensity vibrations such as that from a hand-held vacuum cleaner or engraving tool, or the interruption of sleep by occasional noise of moderate intensity.

The sources of noise are of course very widespread, those giving most concern include factory machines at work, overflying aircraft at home and firearms and pop concerts at leisure.

## Measuring hearing

Ears are tested in a soundproofed room, usually with sound supplied by headphones to each ear in turn. The intensity is gradually increased until the subject indicates that it is heard. The procedure is repeated for selected frequencies to cover the normal range. These hearing thresholds are plotted on an audiogram on which normal hearing is taken to be 0 dB. Figure 3.6(a), shows typical results for people aged 40 and 65, whose loss is due only to ageing (presbycusis). Figure 3.6(b) shows the results for similar age groups who have been exposed to noise with an $L_{eq}$ of 96 dBA for eight hours a day, five days a week since their eighteenth birthday. Note how the higher frequencies are most severely affected.

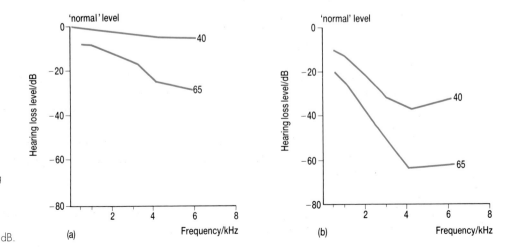

**FIG 3.6** Audiograms of hearing loss;
**(a)** due to ageing (presbycusis)
**(b)** due to exposure to sound level, $L_{eq}$ = 96 dB.

## SUMMARY ASSIGNMENTS

**3.10** Write a brief report on what you consider to be the main unnecessary sources of noise you and your family are exposed to. Give any values of noise levels you can obtain, for example from websites about noise pollution. Explain the possible hazards, and state what actions could be taken to reduce these.

**3.11** Research advice on how to protect against the hazards of listening to loud music. For example from the British Tinnitus Association, or websites, such as that of HEAR for rock musicians. Present this with an explanation of the advice in a form suitable for teenage pop fans.

**3.12** **(a)** Describe briefly the passage of an incoming sound wave from its entry into the external ear canal to entry into the fluid in the inner ear.

**(b)** Figure 3.7 is a graph, with unscaled axes, of the variation with frequency of intensity of sound at the threshold of hearing.

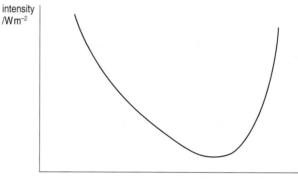

**FIG 3.7**

Draw Figure 3.7 and on the axes, mark stated values, for a person with average hearing, for
**(i)** the audible range of frequencies and the frequency at which the ear is most sensitive,
**(ii)** the threshold intensity of hearing.

**(c)** Some people find a sound frequency of about 2.6 kHz very annoying. This annoyance occurs when the wavelength of the incident sound is four times the effective length of the auditory canal. Estimate, for such a person, the effective length of the canal given that the speed of sound in air is 330 m s$^{-1}$.

**(d)** The intensity of sound at a point distant 0.60 m from a fire alarm is $8.0 \times 10^{-2}$ W m$^{-2}$. Calculate the intensity level of the sound at this point.

*(OCR June 1999)*

**3.13** **(a)** State what is meant by
**(i)** the *intensity I* of a sound,
**(ii)** *threshold intensity $I_0$* of hearing,
**(iii)** *intensity level*.
**(b)** For a person with normal hearing, state values for
**(i)** the threshold intensity,
**(ii)** the threshold intensity level.
**(c)** In some modern audio-amplifiers, the volume control is marked in decibels. Suggest, with reference to the response of the ear, why this is a convenient calibration.
**(d)** A pneumatic drill produces noise of intensity level 98 dB at the ear of a passer-by. Calculate the intensity of this noise at the ear.
**(e)** The intensity level at which discomfort is felt is 108 dB. Calculate the minimum number of pneumatic drills, (identical to that in **(d)** and situated an equal distance from the passer-by) that would cause the passer-by to experience discomfort.

*(UCLES June 1997)*

**3.14** **(a)** The *threshold of hearing* is quoted as $1.0 \times 10^{-12}$ W m$^{-2}$. Explain what is meant by the *threshold of hearing* and state the frequency at which the threshold has this value.

**(b)** Sound intensity levels are usually measured in decibels. Give **two** reasons why this *logarithmic* scale is used.

**(c)** Why was it necessary to introduce an adapted scale referred to as the dBA scale, which is used on some sound level meters?

**(d)** Modern hi-fi equipment and televisions often have volume controls which allow the sound volume to be increased in steps. If each of these steps produces an increase in the sound intensity level of 2.0 dB calculate
  **(i)** the ratio by which the sound intensity is increased for each step up in volume,
  **(ii)** the ratio by which the sound intensity is increased for a total of 10 identical steps up in the volume.

*(AQA June 1999)*

**3.15**

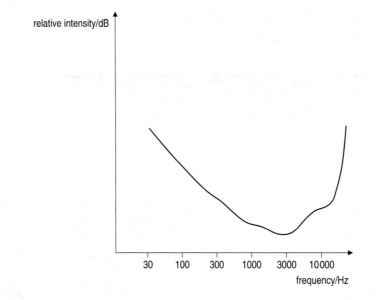

**FIG 3.8**

Figure 3.8 shows an equal loudness curve for a normal human ear at a low level of sound.

**(a)** Describe the procedures to be carried out to obtain an equal loudness curve for a person whose hearing requires testing.

**(b)** Draw Figure 3.8, add, and label clearly, the following additional equal loudness curves:
  **(i)** the curve expected for a normal ear at a much higher sound density,
  **(ii)** the curve expected for an old person whose hearing has deteriorated,
  **(iii)** the curve expected for a person suffering from an industrial noise-induced hearing loss.

**(c)** For each of the additional curves you have drawn on the figure, describe the most significant feature that makes it different from the printed curve.
  **(i)** normal ear (high sound intensity)
  **(ii)** old person
  **(iii)** industrial damage

*(NEAB June 1998)*

# BIOMEDICAL MEASUREMENT

**T**he body contains mechanical, acoustic, thermal, electrical, optical, hydraulic, pneumatic, chemical and other types of system, all interacting with each other. It also contains a powerful computer, several communication systems and many control systems. In many of these systems there is no simple relationship between input and output, and often it is difficult to decide what is the input and which the output, since the body has many feedback loops. The variables to be measured are often inaccessible and often interact with other variables.

So even if the body is considered to be simply a black box, we can see that biomedical measurement is a complex task. The complications are of course significantly greater when the subject of the measurement is a conscious, feeling person. You may be able to identify some of these in investigations you know of – for example dentistry. ■

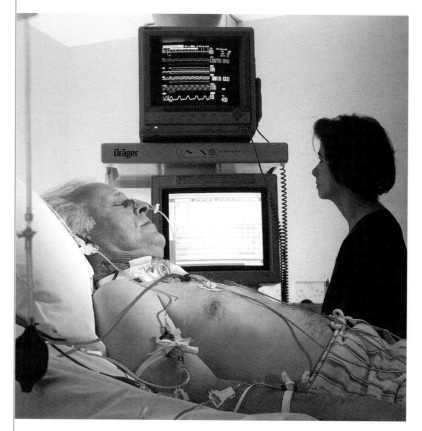

The comprehensive monitoring of a patient's condition in an intensive care unit.

## PREREQUISITES

Before you study this theme you should have some familiarity with:

● The principles of current electricity and its measurement.
● The principles of thermometry, and common thermometers.
● The wave nature of light and sound and their reflection and refraction.
● The functions of the hardware and software in a computer system.

# INSTRUMENTATION

B iomedical instrumentation is the use of instruments and equipment to carry out the task (or part of it!) outlined in the theme introduction; methods of measuring the different systems in the body. In this theme we are concerned only with the most commonly made physical measurements. This chapter will illustrate some of the principles of the process of obtaining and presenting the measurement, using temperature as an example.

## LEARNING OBJECTIVES

*When you have studied this chapter you should be able to:*

① list the purposes of a biomedical instrumentation system and the type of components of which it is composed;

② discuss the general problems involved in measuring human beings, and illustrate these with specific examples;

③ give examples of the uses of active and passive transducers in measuring and the ways in which electrical signals from the transducer can be processed;

④ give brief descriptions of the cathode-ray tube and thermograph, as display devices, and of the technique of telemetry;

⑤ give examples of the use of computer systems in the recording, data processing, transmission and control of measurements.

## 4.1 SYSTEMS

### *Purposes*

The purposes of biomedical instrument systems can be classified as follows:

**Information gathering** will always be a part of the purpose of measurement. In research this may be the full extent of the objective. In any clinical situation other purposes will follow. Instruments for clinical use will typically perform a more limited range of measurements than research instruments, and so be less complex to operate. This enables them to be used routinely by medical staff who are not experts in the equipment, only in the use of the information.

**Diagnosis** of the state of the body, or that part which has been measured, is the most important part of medical practice. In many cases the information is more usefully presented as a qualitative image or picture, rather than a set of quantitative values. Much of the present development of diagnostic instrumentation is concerned with providing both.

**Monitoring and evaluation**. When the condition of the person is known, measurements are made to see if there is any change of this condition as a result of some input, or simply over a period of time. For example, the heartbeat rate and lung capacity of an athlete may be monitored during exercise in a training programme and used to evaluate either her body or the training programme.

**Control**. The measurement may be used to bring some change in the body's performance such as a heart pacemaker to ensure that the heart continues to pump effectively.

## Performance characteristics

These are some of the criteria by which you could judge the suitability and effectiveness of an instrumentation system:

**Range** is the maximum and minimum inputs that the instrument can detect and produce a usable reading from. Inputs might be static, such as the amplitude of a current; or dynamic, such as the frequency of a sound wave.

**Sensitivity** is the smallest variation that can reliably be measured. This determines the resolution of the device, which is the smallest variation that can accurately be read. This is a very important parameter in imaging systems, where it may be dependent on visual contrast.

**Linearity** is the ability of an instrument to be equally sensitive throughout the range over which it operates. Direct current meters that you have used are usually linear; some types of alternating current meters are much less sensitive towards the limits of their range, than in the centre.

**Frequency response**. Many signals from the body are rapidly varying, so the instrument needs to be sensitive over the range of frequencies encountered.

**Signal-to-noise ratio** determines the ability to distinguish the required measurement (the signal), from any background or other effects (the noise). Clearly the sensitivity of an instrument is greater when the signal-to-noise ratio is low. The sources of noise are electrical interference, for example from 'mains hum', and to a lesser extent, thermal noise which arises from the nature of the electrical circuit itself.

**Isolation**. Safety standards require that measurements must be made so that there is no direct electrical connection between the subject and the ground or other equipment. **Telemetry** is one way of achieving this in which the measuring device in or on the body, transmits the information by radio waves.

Other important criteria which apply to any measuring instrument are, **accuracy** (which will include considering if the presence of the instrument affects what it is measuring), stability, and last but not least, simplicity.

Mains electricity in Britain is a.c. with a frequency of 50 Hz. As it flows in a cable it will generate an accompanying magnetic field. Any conductor near to a mains-carrying cable will have a current induced in it by this, which behaves as a noise to any signal in the conductor. The current changes direction 100 times per second, so the noise has a frequency of 100 Hz. This is known as mains hum.

## INVESTIGATION

### Measuring body temperature with a thermocouple

We will use this simple experimental procedure to examine aspects of the instrumentation system described above.

### Procedure

1. Obtain or make a simple thermocouple by joining together dissimilar wires such as copper and constantan (details may be obtained from a standard physics practical book).

2. A reference junction should be kept at 0 °C and the other junction is placed where the temperature is to be measured. Choose suitable body sites.

3. The temperature difference of the junctions generates an e.m.f. of about one millivolt. This is measured by a sensitive voltmeter or potentiometer. Choose a suitable instrument with a high impedance.

4. The voltage reading represents a certain temperature. Tables are available to give this, or alternatively you can calibrate the instrument against another thermometer.

### Suggestions for investigation

1. You may need to try out methods of improving thermal contact between thermocouple and skin, for example by interposing a copper block.

2. What is the effect of hot and cold water, how quickly does the skin respond?

3. What is the effect of a fan on wet and dry skin?

4. What is the effect of warm clothing?

5. Does body temperature vary with the time of day?

## QUESTIONS

After you have carried out the above procedure and measured your own body temperature, answer these questions about the instrumentation system that you used.

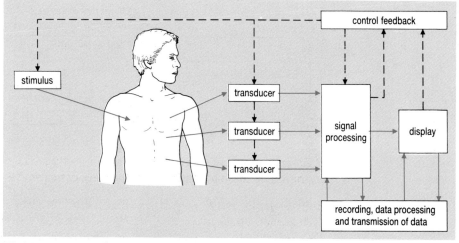

**FIG 4.1** Components of an instrumentation system.

**4.1**  How good were the performance characteristics of your system? Comment on each of the headings listed above, which are relevant to your measurement. For example,

   **(a)** were the range and sensitivity of the instrument suitable for this use? If not, how could these characteristics be improved?

   **(b)** What are the sources of error in your measurement, with what accuracy do you give your result?

   **(c)** What affects the stability of your instrument?

   **(d)** Your instrument was probably quite simple. To what extent do your measurements rely on reference to a more complex or quality-controlled instrument?

**4.2** (a) Draw a labelled block diagram similar to Figure 4.1, to show the components of your system. Write a brief description of each of the **components** you used, e.g. stimulus-ambient temperature and the particular conditions you investigated.

(b) How would you include a feedback loop, for example to prevent a patient with a fever from overheating?

The core temperature of the body, or systemic temperature, is maintained by careful control of the energy lost to the environment. Its value measured by mouth is about 37 °C, or one degree lower under the armpit, or one degree higher in the rectum. Body temperature of even a healthy person, varies by about 1 °C over a daily cycle, being lower earlier in the day. This temperature is affected temporarily by exercise, over longer periods by illness, but not by ambient temperature. The temperature control centre is located deep inside the brain. It monitors the blood temperature and controls various parts of the body in order to increase or decrease the rate of energy release. By contrast the skin temperature varies, usually between about 30 °C and 35 °C, as a result of local factors such as the exposure to ambient temperature, the amount of fat covering the capillaries and the blood circulation through them. Table 4.1 gives some indication of how intolerant the human body is, of changes in body temperature.

**TABLE 4.1** Effect of temperature change on the body

| CORE TEMPERATURE/°C | CHANGE | CONDITION OF SUBJECT |
|---|---|---|
| 45 | hyperthermia | |
| 44 | | Dead |
| 43 | | |
| 42 | | Central nervous system affected; |
| 41 | | convulsions |
| 40 | | Dilation of peripheral blood vessels, increased |
| 39 | | heart rate, reduced flow of blood to brain |
| 38 | | |
| 37 | | Normal |
| 36 | | |
| 35 | | Constriction of peripheral blood vessels, |
| 34 | | reduced heart rate; shivering |
| 33 | hypothermia | Depression of nervous system function; |
| 32 | | amnesia |
| 31 | | |
| 30 | | Temperature regulating system fails, sleepiness |
| 29 | | depression of respiration, heart beat affected, |
| | | loss of consciousness |
| 28 | | |
| 27 | | Dead |
| 26 | | |

## QUESTION

**4.3** The measurement of body temperature is a routine check to monitor the overall condition of someone who is unwell. It is simple but suffers from several limitations which originate from the subject rather than the instrument. Identify as many as you can, and comment on how serious they are.

## CASE STUDY

### How a fun run meant meltdown for Mark Dorrity's body

By Sue-Ellen O'Grady

On February 27 this year, Mark Dorrity set off on what he expected to be an easy 8-kilometre fun run in Wagga, southern NSW. But near the finishing line, the fit 28-year-old collapsed, his body destroyed.

In less than an hour, his thigh muscles had overheated, liquefied and died. One leg has since had to be amputated at the buttock, because of gangrene.

Before Mark collapsed, his kidneys failed because the dying muscles had released toxic proteins into his blood, which thickened to a molasses-like consistency. Every organ in his body was affected.

He suffered brain damage. His lungs could not function unaided. His buttock and hamstring muscles also liquefied, but not as severely as his thigh muscles. Mark's heart stopped at least once. When it started again, it hammered away at 150 beats a minute, compared to its normal 70. He was on a dialysis machine for eight weeks, and in a coma for three months. When he regained consciousness, he could not walk or talk.

Even now, five months later, Mark cannot turn over or get out of bed unaided. He faces months of intensive rehabilitation.

The devastating damage to Mark Dorrity's body was caused by heat exhaustion and dehydration resulting in a rare condition known as rhabdomyolysis, the extreme result of what every runner and athlete knows as muscle fatigue.

The director of research at the Sports Medicine Institute, Dr Tony Miller, says the condition usually affects runners taking on more than they are used to in training.

'In Mark's case it was caused by his body severely overheating – to 42.8 degrees. At the same time, he was extremely dehydrated. When someone has a temperature that high, they are delirious. They ignore the body's warnings to stop.'

Mark Dorrity was no weekend jogger. When he graduated from the University of NSW with an honours degree in Science in 1974, he won a Blue for athletics. He moved to Melbourne to work as a wool exporter, and ran four kilometres through the Botanical Gardens every day. As well, he swam a kilometre three times a week. He had minimal body fat.

He travelled to Wagga in February with a group of friends, all planning to compete in a local event. When the temperature that day rose to 42 degrees, the locals cancelled their run.

But Mark and his friends, deceived by the dry heat, decided to hold their own race.

'It just didn't feel that hot,' he remembers. 'So we ran off.'

He drank several glasses of water before beginning to run, but none during the race. That, say doctors, proved to be his near-fatal mistake.

When Mark collapsed, he was leading the race by a kilometre. Friends driving alongside rushed him to the local hospital where he was packed in ice to lower his body temperature. He remained there for two weeks before being moved to St Vincent's for dialysis treatment.

Mark recalls nothing of this 'I don't recall collapsing. I remember waking up twice at Wagga Hospital, and then waking here.'

'How do I feel? I'm very lucky to be alive. I know that I'm a medical miracle. And it's a warning to other runners to be extremely careful.'

*The Sydney Morning Herald* 3.8.1988

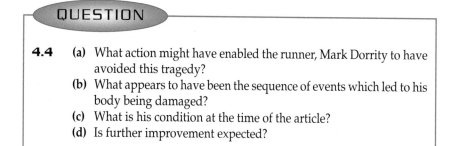

## QUESTION

**4.4**   (a) What action might have enabled the runner, Mark Dorrity to have avoided this tragedy?
   (b) What appears to have been the sequence of events which led to his body being damaged?
   (c) What is his condition at the time of the article?
   (d) Is further improvement expected?

## Problems of measuring a living subject

The answer to question 4.3 indicates some of the difficulties in a simple body measurement. Many of the variables that the clinician wants to know about are much less accessible than temperature, and subject to many unavoidable influences, both known and unknown. The presence of the transducer may physically affect what it is measuring, for example by obstructing the flow of blood. It may also have a psychological effect which can affect the physiology being measured, through fear. Many readings and images are altered just by the movement of the patient. This can give false results called **artefacts**. A general problem which is particularly serious in radiological measurement is the effect of the energy applied to the body during measurement. In the extreme this can cause more damage to the patient than the disease that is being investigated. The issue of safe practice is always of prime importance, and may impose severe restrictions on the use of high voltages and high intensity radiations.

Most of the measurements described in this book take place in the body, this is referred to as **in vivo**. This reflects much medical practice, though there are tests which are done on samples taken from the body. These are called **in vitro**, (in glass) tests, and in some cases this can overcome the problems of measuring the body directly.

## 4.2 TRANSDUCERS

**T**hese are the devices that change the physiological variables to be measured into electrical signals. They are part of a circuit which produces an electrical output from a non-electrical input. The relationship between input and output need not be linear but must be unambiguous; that is, for any value of the input there is only one value of the output. Some transducers actually convert the input energy into electrical energy, these are called **active** transducers and the thermocouple is one example, see Figure 4.2(a). In others there is simply an electrical change produced in the circuit of which they are a part, for example the resistance of a conductor will change if its dimensions change. This is made use of in detecting movement or pressure with a strain gauge. This is a **passive** transducer see Figure 4.2(b). In practice it is not significant to which group a transducer belongs, except that some active conversions are reversible, whereas passive effects are not. This can be important for example in ultrasound where the transducer both generates and detects the signal. Table 4.2 lists some basic transducers used in biomedical applications. A brief comment on each follows.

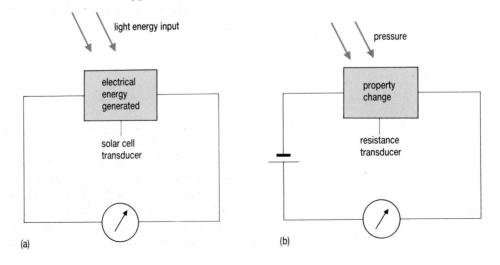

**FIG 4.2** Transducer types
**(a)** active
**(b)** passive.

**TABLE 4.2** Transducers for biomedical applications

| PHYSICAL VARIABLE | TYPE OF TRANSDUCER |
| --- | --- |
| Force (or pressure) | Piezoelectric strain gauge |
| Displacement | Variable resistance |
| | Variable capacitance |
| | Variable inductance |
| Surface strain | Strain gauge |
| Velocity | Magnetic induction |
| | Doppler shift |
| Light | Photovoltaic |
| | Photoresistive |
| | Photodiode |
| Magnetic field | Hall effect |
| Temperature | Thermocouple |
| | Thermistor |

The **piezoelectric effect** is the conversion of movement energy into an electric current, and vice versa. It is used in both directions in ultrasonics and is described in Chapter 8.

Movement or displacement can be detected by the movement of (a) a slide wire to change the length of **resistance** wire in a circuit, as in a potentiometer, (b) a magnetic core in a coil or a transformer to produce a change in **inductance**, (c) the plates or dielectric of a capacitor to change the **capacitance**. If the movement is more rapid it can be used to produce **magnetic induction** by the relative movement of a coil and a magnet. **Strain gauges** detect a range of mechanical changes and these are described in Chapter 6.

The **photoelectric effect** is used in the silicon **photovoltaic** cell or photocell which is the basis of the charge coupled device used for video endoscopy, described in Chapter 7. In semiconductor materials such as cadmium sulphide the conductivity is increased by the absorption of light. This is the **photoresistive** cell which is a sensitive transducer with a rather limited frequency response. The **photodiode** is a semiconductor transducer which is less sensitive but with a better frequency response.

### QUESTION

**4.5** Make a list of the body systems or functions that you can think of to measure, with the transducers listed in Table 4.2. (You may find it helpful to discuss this with a fellow student and then pool your ideas.)

## Temperature transducers

In most situations the clinical mercury thermometer is the standard instrument. This is an example of a transducer which does not produce an electrical output. If continuous recording of temperature is required then a **thermocouple** or a **thermistor** is used. The latter is now more common because it has a greater sensitivity and does not require a reference junction.

A typical thermistor for biomedical use is shown in Figure 4.3(a). It consists of a bead of semiconductor material, usually a mixture of nickel, cobalt and manganese oxides, enclosed in a thin glass envelope and attached to connecting wires. The sensitivity is typically ten times that of a thermocouple, enabling temperature differences of as little as 0.01 °C to be measured. The disadvantages are that thermistors are more expensive and have a shorter life than metal thermocouples as they are susceptible to overheating, and their temperature coefficient of resistance is exponential (and negative) which means they require complex circuitry to be used as signal measuring devices. An example for laboratory use is shown in Figure 4.3(b). There are thermistors with a positive coefficient of resistance, called positors, but these are not in common use.

semiconductor

glass
envelope

0.1 mm

leads

(a)                                                                (b)

**FIG 4.3** Temperature detectors **(a)** small thermistor probe **(b)** laboratory thermistor probe.

Newer thermoelectric devices have been produced with very small dimensions and fast response times, and automatic cold junction compensation is incorporated in the circuit. A number of thermocouples can be connected together to produce a greater sensitivity, at the expense of increased size. Such a device is called a **thermopile** and it is used to detect the heat radiated from the skin. The relationship between energy radiated ($E$/joule) and temperature ($T$/kelvin) is given by **Stefan's Law**:

$$E \propto \sigma A T^4 \qquad (1)$$

where $A$ is the surface area and $\sigma$ is Stefan's constant. The thermopile is usually calibrated to read temperature directly. This last effect can be used to detect defects in the circulatory system by recording the skin radiation, as described in a later section.

## 4.3 DISPLAY DEVICES

**A** great variety of instruments is used to present the electrical outputs of transducers in a form that the operator can use. It is obviously important that these are suitably matched, for example it would have been no use trying to measure the output of your thermocouple with an ammeter. Processing circuits such as amplifiers, pulse counters and filters are introduced to provide an appropriate input to the display device, those specific to a certain measurement are described in the relevant chapter. We will not be describing electrical meters in this book – they are widely used in biomedical measurement, but in principle they are identical to the meters you will already have studied and used.

For medical diagnosis the most useful form of display is often a picture of the organ or system being investigated. This means that signals have to be located in a certain position and reproduced in that position on a screen. This may be a simple process of illuminating a film, as in the case of X-radiography, or it may involve complex scanning procedures, as with X-ray tomography. For immediate, so-called 'real-time' viewing the cathode-ray tube is invaluable, and very widely used.

# Cathode ray tube (CRT)

As a display device the important features of the CRT are its ability to reproduce electrical information in three dimensions, vertical, horizontal and brightness or colour, and so build up a picture.

The details and operation of the laboratory CRT will be found in a standard physics course textbook. A typical example is the display of a rapidly varying electrical voltage, say the output of a microphone. This signal is fed to the Y-deflection plates, and produces a vertical displacement in proportion to the amplitude of the signal. At the same time a regularly varying voltage called the **time base** is applied to the X-deflection plates, so that the change of the output is displayed with respect to time as the horizontal axis, see Figure 4.4(a). This type of display uses only the vertical and horizontal dimensions and is quite common in medicine. You will find examples in Chapter 5 (the electrocardiogram, ECG), and Chapter 8 (the ultrasound A-scan).

All three dimensions are used in television, where time base voltages are applied to both the X- and Y-deflectors so that the electron beam scans the whole surface of the screen, at a rate which gives a continuous image.

The signal is fed to the electron gun so that it controls the intensity of the electron beam and hence the brightness (and colour) of the image produced at a particular position on the screen. By linking the scanning of the screen with the scanning of the subject by the camera, the brightness distribution forms a picture of the subject, see Figure 4.4(b). Examples of this are to be found in all of the later chapters of this book, where the particular features of the system for that type of signal (ultrasound, X-ray, radioisotope etc.), are described, that for infrared is described below.

The use of computers for storage, processing and retrieval of information has extended the use of CRTs or monitors for display.

In the 625 television system used in Britain, the vertical time base moves the beam down the screen once every 1/50 second, while the horizontal time base moves the beam across the screen tracing out 312.5 lines. In the next 1/50 s another 312.5 lines are traced out, so that in 1/25 s 625 lines are traced out.

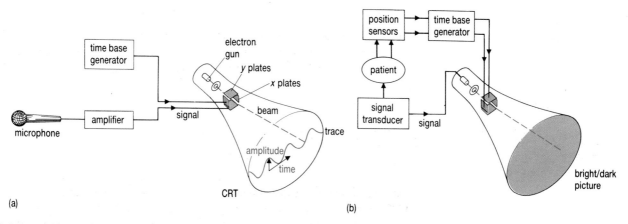

(a)  (b)

**FIG 4.4** Cathode ray tube displays **(a)** amplitude/time **(b)** brightness/scan.

# Thermography

This is a technique to display images of the infrared radiation from the skin. Stefan's law (equation 1) can be written as

$$E = e\,\sigma\,AT^4$$

where $e$ is the **emissivity**, a constant which depends on the nature of the surface. Black and matt surfaces which are good absorbers and emitters have $e \sim 1$, whereas shiny surfaces have a low value of $e$. A perfect radiator with $e = 1$ is called a black body. The emissivity varies somewhat with wavelength. In the visible range (e.g.

sunlight) the darkest human skin has an emissivity of 0.82 and the lightest 0.65. Bodies at room temperature will radiate in the infrared region, for which all human skin has an emissivity of almost 1.

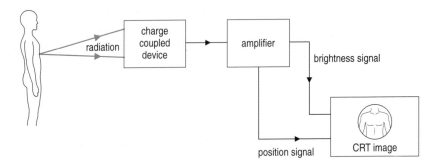

**FIG 4.5** A schematic diagram of a thermographic unit.

The basic system is shown in Figure 4.5. The radiation from the selected area of the subject's skin falls on to the detector. The modern detector is usually a **charge coupled device.** This is an array of electrodes in which charge is generated when infrared radiation falls on it. Thus a current is produced in proportion to the radiation. A two-dimensional picture is built up on a CRT screen (see section 7.2 page 101 for details). The brightness or colour of the visible image is related to the temperature of the radiating surface.

The resulting display can indicate abnormalities in the skin temperature, and therefore in the underlying tissue. Thermography is used to detect direct circulation problems such as those occurring in diabetes, gangrene and vascular disease. This can enable preventative measures to be taken to avoid complete loss of circulation and the amputation of a limb. Tumours produce a local change in the circulation system and so this method can be used to detect testicular cancer.

**QUESTION**

**4.6**  Figure 4.6 shows that people get hot playing squash! What more can you tell from these thermograms?

**FIG 4.6** Thermogram of a man **(a)** before playing a game of squash and **(b)** immediately afterwards.

**T**he principal method of transmission in any biomedical measurement system is electrical. The transducer converts the signal to be measured into an electrical one which can then be amplified, digitised, or in other ways processed so that it is suitable for transmission, either to other parts of the system, or to a different place. There are however a number of situations in which direct electrical connections are either impossible or undesirable. Three examples are the monitoring of mothers in labour, astronauts in space and cardiac patients during recovery. In these circumstances a method of transmitting signals over distance is used. This is called biotelemetry.

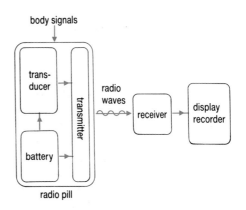

**FIG 4.7** Radiotelemetry system.

## Biotelemetry

One of the simplest examples of taking readings at a distance is the stethoscope. The sound of the heartbeat is amplified acoustically and transmitted along a tube to the ear of the physician. The transmission of the electrical signal from the heart is probably the most widespread use of telemetry, at the present time. It was also one of the earliest, having been transmitted on the telephone system by the originator of the electrocardiogram in 1903, (see Chapter 5). Telemetry's common use nowadays is between the place of a cardiac emergency such as an ambulance and the hospital, where a cardiologist can interpret the signal and instruct the ambulance crew in resuscitation procedures. The method of communication is usually radio, and the methods are called **radiotelemetry**. Other frequent uses are in monitoring patients who have certain conditions which put them at risk. In such cases it may be necessary to implant the assembly of transducer, transmitter and power supply. This so-called 'radio pill' (Figure 4.7), is either swallowed or introduced surgically. The radio transmitter is essentially an electronic circuit which produces an alternating current (the oscillator) in a coil of wire (the aerial). This generates a radio wave which is detected in a corresponding coil, the receiver.

Most of the important parameters of the body have been measured and transmitted in this way, using appropriate transducers in small, encapsulated pills to protect them against body fluids. Examples include pressures and temperatures in various body cavities such as the vascular system, stomach and the intestines, and the pH of the stomach and oesophagus.

**T**he availability of ever greater computing power at reducing cost in recent years has affected hospitals, as much as anywhere else. The use of computers in measurement, imaging and control is now very widespread, and has transformed the types of equipment now in use as the following chapters will show. The development of databases for patient records and so on is now changing the way hospitals are run.

## Components of a computer

Essentially a computer is an electronic device which can automatically and rapidly perform a long and complicated sequence of operations as directed by instructions called a *program*. It can also store and retrieve large quantities of information. All computers comprise four elements: the *calculator* or *arithmetic unit*, the *memory*, to store data and instructions, *input-output (I/O) devices*, to allow the computer to communicate with the world, and the *control unit* (Figure 4.8).

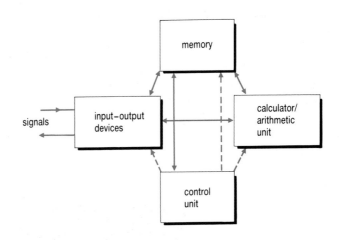

**FIG 4.8** Computer elements.

The reason for the rapid growth of computing has been the successful miniaturisation of electronic circuits in semiconductors. This has led to the production of tiny integrated circuits, microprocessor units and memories of great complexity, at relatively small expense. The basic structure of all these components is like a series of switches, and the fundamental operation is that of switching 'on' or 'off'. All information processed by the computer must be reducible to this binary state, when it is called a *bit* of information. These characters are often manipulated in groups of 32 or 64 bit words, called *bytes*. Most information from biomedical measurement is produced as a continuously varying, *analogue* signal, so it must be changed to *digital* information, that is given a numerical value, before the computer can handle it. This is done by an *analogue-to-digital converter* (A/D), and its reverse, *the digital-to-analogue converter* (D/A), is used to display the product of the computer processing. This electronic circuitry and physical equipment is called the *hardware*; the programs of instructions, the computer *software*.

This is of two types, the *system* software which manages the operations of the computer and the *applications* software which enables specific functions to be carried out.

# Biomedical applications

Examples of specific computer uses will be found throughout the remainder of this book. They include the following categories:

- data acquisition,
- data processing,
- mathematical calculations,
- pattern recognition and limit detection in monitoring,
- storage and retrieval,
- statistical analysis,
- data presentation,
- control.

## INVESTIGATION

### Temperature monitoring using sensors and a computer

Consider how the simple measurements of body temperature that you made earlier in this chapter could be extended to investigate the dangers of athletes overheating, as in the news item about Mark Dorrity.

1. Plan an investigation incorporating the following features and any others that you consider important:
   (a) continual measurement of temperature whilst the exercise is being performed;

(b) a method of warning the athlete when his/her temperature rises too high;

(c) the effect of water consumption on temperature.

2. Discuss your plans with your teacher and carry out as much as possible depending on the datalogging equipment and time available to you

3. Record your results and communicate any conclusions you may draw from the work, in an appropriate form, e.g. to the P.E. department of your school or college.

# Whole-body scanning for medicine

**FIG 4.9** Whole-body scan.

A technique used in the clothing industry for accurately measuring people's shape could be equally useful in medicine. A team at University College London is refining the technique and exploring the possibilities – which could range from individualised prosthetics to more accurate ways of calculating drug dosages for children.

The equipment consists of eight cameras mounted on a U-shape frame in the walls of the cubicle. Using harmless near-infrared radiation, the cameras scan the subject, moving from the head to the feet. In each 10-second scan, the cameras take 400 'slices' through the body, creating an initial data set of more than 100 000 points.

Software uses these to construct the body surface shape and make calculations such as body volume and surface area (Figure 4.9).

**SUMMARY ASSIGNMENTS**

**4.7** Write an essay on the increasing use of computers in biomedical measurement. Use the description of the components and the list of application categories, together with some background reading of current journals, books and articles from the bibliography, to ensure that your account is up-to-date and clearly explained.

*(You may wish to leave this question until you have studied in more detail, some specific areas of measurement and imaging in later chapters.)*

**4.8** The stages in a transducer system may be represented by the boxes shown below.

    **(i)** Give **one** example from medical physics of a measurement system in which a transducer measures a physical quantity.

    **(ii)** For your example state what would be the input signal from the patient, the transducer, and the display device.

*(NEAB June 1998)*

**4.9** The simplified circuit diagram (Figure 4.10) shows a thermocouple which could be used to measure the temperature of a patient in hospital.

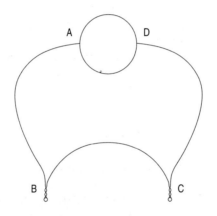

**FIG 4.10**

    **(a)** **(i)** Draw and label each of the three wire parts, AB, BC and CD with a suitable metal.

        **(ii)** Fully specify a suitable measuring instrument to be connected between A and D.

    **(b)** State **one** significant advantage and **one** disadvantage of monitoring the temperature of a patient in a hospital by using a thermocouple.

    **(c)** Thermocouples depend on an effect which generates an e.m.f. State the factors which determine the magnitude of this e.m.f.

*(NEAB Mar 1998)*

**4.10** An image of the human body can be obtained by a thermographic scanning process.

    **(i)** State the type of radiation detected from the body which produces the image.

    **(ii)** What information is provided by the intensity or false colour of the image?

    **(iii)** Name **one** medical application of this imaging technique.

    **(iv)** State an important limitation of this technique.

*(NEAB Mar 1997)*

# 5

# ELECTRICAL SIGNALS

There are many important systems in the body which generate electrical signals in the course of carrying out their functions. These are the bioelectric potentials associated with nerve conduction, heartbeat, brain and muscle activity, and so on. By the use of electrodes these signals can be measured and used in medical diagnosis. This chapter is an account of the process, with particular reference to that most vital of organs, the heart. The therapeutic use of electricity applied to the body is also reviewed.

## LEARNING OBJECTIVES

*When you have studied this chapter you should be able to:*

①   give an illustrated account of the origin of biopotentials in the human nervous system;

②   describe the action of the heart as a pump to the circulation system;

③   describe the method of measuring an electrocardiogram (ECG), recall its main features, and explain its use in diagnosis;

④   give examples of other measurements of the body's electrical activity, and of the therapeutic use of electricity.

## 5.1  BIOPOTENTIALS

Luigi Galvani, an Italian anatomy professor, discovered the presence of electricity in living tissue, with his observations, in 1786, of the stimulation of a frog's leg, which you may yourself have seen repeated. It was not until 1903 when the Dutch physician Willem Einthoven found a way of measuring these potentials, that any practical use was made of them. The advent of microelectronics and the computer have now made their measurement a simple routine, and led to important advances in knowledge of the functioning of the body.

## Nerve cells

The communication system of the body is the nervous system, which passes its messages along long, thin nerve fibres, called *axons*. A typical nerve which controls movement is shown in Figure 5.1. The main cell body, containing the nucleus is situated in the spinal column, and the axon runs from here to the muscle to be moved. So the axons to the foot for example, are up to 1 m long. The nerves pass messages between each other at the *synapses*, which are connected to the branching ends of the nerves, the *dendrites*. The axon is typically only a few micrometres in diameter, and covered with a sheath of *myelin*. This acts as an insulator between the axon and the surrounding tissue. There are gaps in this sheath every millimetre and at these, the *nodes of Ranvier*, charge can flow, by the passage of ions, to attenuate (reduce), or amplify (increase) the nerve signal. This flow will depend on the local biopotential.

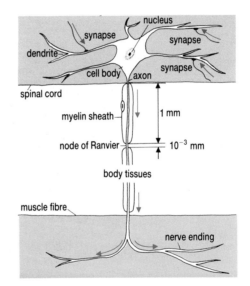

**FIG 5.1** A nerve cell from spinal cord to muscle.

## Resting and action potentials

The axon wall is a membrane through which only certain things can pass under certain conditions. The small $K^+$, $Na^+$ and $Cl^-$, ions can permeate but not the larger organic ions such as proteins, so the membrane is said to be semipermeable. In its resting state, the fluid inside the axon, the axoplasm, contains an excess of $K^+$ and negatively charged organic ions, and the fluid outside, the interstitial fluid, contains an excess of $Na^+$ and $Cl^-$ ions. The imbalance between the $Na^+$ and $K^+$ ions is maintained by an imperfectly understood process called the sodium-potassium pump. This transfers $Na^+$ out of the cell and $K^+$ into it. Because the membrane is relatively impermeable to $Na^+$ ions, they do not diffuse back with the concentration gradient, but the $K^+$ ions can. This results in an excess of positive charge on the outside and a potential difference between the inside and outside of about 90 mV, the inside being negative compared to the outside (Figure 5.2). This is the **resting potential** of the axon, which is said to be polarised.

**FIG 5.2** The source of biopotential in a nerve fibre.

**FIG 5.3** Resting and action potential of a nerve.

When there is a stimulus to the nerve the membrane suddenly becomes much more permeable to the Na⁺ ions, which therefore flow into the axon to reduce the potential difference. This process of depolarisation takes only a few milliseconds, and continues until there is a positive potential inside of about 20 mV (Figure 5.3). This is called the **action potential**. As soon as this potential is reached, the membrane again becomes less permeable to Na⁺ and more permeable to K⁺ which flow out of the axon to repolarise the cell. Over a longer period of time (about 50 ms) the Na⁺ ions and the K⁺ ions are exchanged across the cell wall, by the sodium-potassium pump. This ensures that the resting axon is in a state that can respond to the next stimulus.

It is called a pump because it builds up an energy store in the imbalance of the ions across the membrane, like water being stored behind a dam. When the stimulus causes the permeability to change, the flow of charge is like the water flow when the floodgates are opened. The whole operation is designed to amplify the electrical stimulus signal as it passes along the nerve fibre at a speed of about 150 ms⁻¹ with the depolarised region triggering a change in the adjacent section of the axon.

The changes in the potential are a characteristic of the nerves concerned, those in the heart usually travel more slowly, last longer (150 to 300 milliseconds) and rise to a higher level (40 mV), (see Figure 5.6).

### QUESTION

**5.1** Draw a diagram to show the movement of ions to match the sequence of these changes: resting state – depolarisation – reverse polarisation – repolarisation.

## 5.2  THE CARDIOVASCULAR SYSTEM

The circulation of the blood by the heart is the most important system in the body, if this fails to work satisfactorily death occurs in a very short time. It is not surprising therefore that measurements of this system are routine. In this chapter we will cover the use of biopotentials in the operation of heart muscles, in the next, the measurement of blood flow.

## Components of the system

The heart (*cardio-*) can be considered as a double pump, circulating the blood through two major tubular (*vascular*) systems, the *pulmonary* circulation to the lungs, and the *systemic* circulation to the rest of the body. A simplified anatomical diagram of the heart is shown in Figure 5.4 and a schematic illustration and description of the system in Figure 5.5.

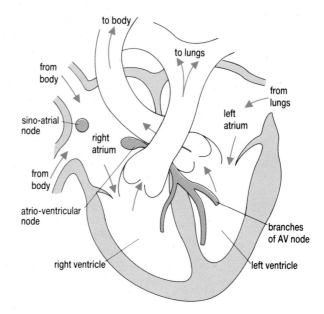

**FIG 5.4** Cross-section of the heart, from the front.

**FIG 5.5** Schematic diagram of the circulatory system.

**QUESTION**

**5.2**  (a) Make brief notes on the cardiovascular system from the above information, if you have not previously studied it (supplemented by a GCSE biology text if necessary).
(b) Why would you expect the walls of the left ventricle to be much stronger than those of the rest of the heart?
(c) Why is the blood that leaves the heart in the arteries at a much higher pressure than that which returns to it via the veins? Why is the pressure in the pulmonary artery less than that in the aorta?

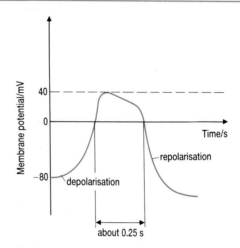

**FIG 5.6** The action potential of the heart.

## Controlling the heart's action

All muscular activity is associated with the changes in electrical potentials, which result from the migration of ions, as described above. The regular pumping action of the heart is controlled by special muscle cells, the *sinoatrial (SA) node* or *pacemaker*, located in the right atrium (Figure 5.4). This produces an electrical stimulus about 70 times a minute and initiates the depolarisation of the nerves and muscles of both atria. These contract and pump blood through one-way valves (tricuspid and mitral) into the ventricles. Repolarisation follows as the ions move to reduce the potential, and the muscles relax. The electrical signal then passes to the *atrioventricular (AV) node*, which initiates the depolarisation of the two ventricles. They contract and pump blood through two more one-way valves (pulmonary and aortic) into the two circulatory systems. The ventricle nerves and muscles then repolarise and relax so that the sequence starts again. The action potential is shown in Figure 5.6, and the complete cycle represents a heartbeat. The rate will depend on the body's demands on the circulatory system, being greater for example at a time of greater physical exertion.

## **5.3** ELECTROCARDIOGRAPHY

### The electrocardiogram

During each heartbeat, the spread of the action potential through the heart, causes potential differences between depolarised and polarised cells. The conducting nature of body fluids transmits these potential differences so that they can be detected at the surface by electrodes. The signals are very much reduced in size by their passage through the body having typical values of 1 to 2 mV. They need amplifying before they can be recorded. Electrodes must be placed in particular positions, and the patient must be relaxed so that no other signals interfere with the heartbeat. Their display over time is called an **electrocardiogram (ECG)**, and their study, electrocardiography, is a method of diagnosing various diseases and conditions of the heart.

A typical ECG is shown in Figure 5.7 and includes the following important features, which are usually referred to by letter labels:

■ The P-wave due to the depolarisation and contraction of the atria.

■ The QRS-wave due to the depolarisation and contraction of the ventricles.

■ The T-wave due to the repolarisation and relaxation of the ventricles.

**FIG 5.7** An ECG trace.

> ### QUESTION
>
> **5.3**  With reference to Fig 5.7,
>   **(a)** Why is the wave for the repolarisation and relaxation of the atria not visible?
>   **(b)** What percentage of the original membrane potential is detected at the surface (refer to Fig 5.6)?

## Measuring the ECG

This was first carried out by Einthoven in 1903. Though the equipment has changed out of all recognition since that time (not least due to considerations of electrical safety! (Figure 5.8), much of the terminology which originated then has become standard throughout the world.

The waveform measured at the surface depends not only on the subject, but also on the placing of the electrodes. At least two will be needed to measure a difference. These are called *bipolar leads*, and any two of three limbs are used. The right leg is not used as it is farthest from the heart, though it may be used as an earth connection to minimise interference from the equipment. In *unipolar leads* one or more electrodes are held at zero potential, while the other records differences. Thus

**FIG 5.8** The first electrocardiograph machine, Cambridge 1908.

one of these limb sites can be compared to the average potential of the other two, or all three limb sites can be connected together to give a neutral, with the 'active' electrode placed in one of six specified chest sites (Table 5.1). This gives a total of twelve alternative arrangements. It is obviously most important in any diagnostic use to know exactly where these were situated for a given ECG, so a labelling system of these standard sites is used, as shown in table 5.1.

**TABLE 5.1** Labelling of ECG leads

| TYPE OF LOAD | ACTIVE ELECTRODE | NEUTRAL ELECTRODE(S) | NAME |
|---|---|---|---|
| Bipolar | Right arm | Left arm | Lead I |
| | Right arm | Left leg | Lead II |
| | Left arm | Left leg | Lead III |
| Unipolar limb | Right arm | Left arm and leg | aVR |
| | Left arm | Left leg and right arm | aVL |
| | Left leg | Right and left arm | aVF |
| Unipolar chest | One of six chest sites | Both arms and left leg | $V_1$ to $V_6$ |

## INVESTIGATION

### *Measurement of ECG*

There are a number of commercially available sets of equipment for measuring and displaying ECGs. If one is available to you, take this opportunity to carry out a simple investigation, following the manufacturers' instructions.

1. Try the different electrode positions and compare results.

2. What is the effect of exercise, the time of day, a stimulant such as coffee etc. on heart rate?

QUESTIONS

**5.4** Describe the results of your investigation, and discuss the factors which were important in the production of a reliable, reproducible signal. (If you were unable to carry out a practical investigation you should still be able to make some predictions!)

**5.5** What are the advantages of the following presentation styles of ECGs and when would each be most useful:
**(a)** continuous monitoring by display on a CRT screen;
**(b)** a permanent record (photographic or numerical)?

**5.6** **(a)** Fig 5.9 shows a modern ECG recorder. What are the advantages of the various features given in the caption?

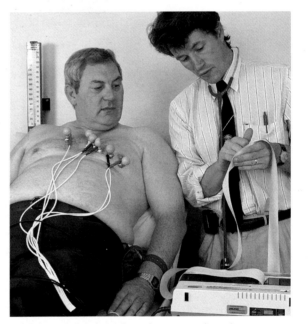

**FIG 5.9** A modern three channel ECG recorder with store, playback and automatic diagnosis facilities.

**(b)** Work out the pulse rates (in min⁻¹) from the ECGs of Fig 5.10.

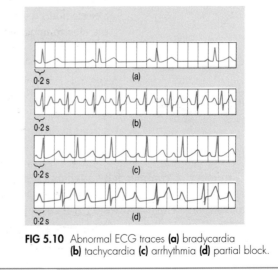

**FIG 5.10** Abnormal ECG traces **(a)** bradycardia **(b)** tachycardia **(c)** arrhythmia **(d)** partial block.

## Diagnosis

**Heart attack**, myocardial infarct, and coronary thrombosis are names given to the blockage of a coronary artery, depriving the heart muscle of blood. The effect is a sudden collapse by the person who suffers severe chest pains. By the age of 65, one man in five in Britain has had a heart attack; about 50 per cent of them will have been fatal. Total deaths are about 150 000 a year making this the most frequent cause of death.

The following are examples of the indications of cardiac disorders that can be detected with the ECG:

- Normal heart rate is 60 to 100 beats per minute. A slower rate is called *bradycardia*, Figure 5.10(a), which may occur in athletes. A faster rate may be produced by sudden emotion, exercise or fever, and is called *tachycardia*, Figure 5.10(b).

- Irregular trace repetitions indicate irregular pumping, *arrhythmia*, Figure 5.10(c). This is quite common in young people. It is increased by deep breathing and reduced by exercise.

- A blockage of part of the heart, due to heart disease will result in a part of the trace not showing. If the heart muscle has been damaged, for example by a 'heart attack', the pumping action will be defective and the wave heights reduced, Figure 5.10(d).

- If the heart fails to pump properly, for example as a result of shock, this will show as a very irregular trace. The condition is called *ventricular fibrillation*, and it can lead to death in a very short time.

# OTHER ELECTRICAL POTENTIAL MEASUREMENTS

**5.4**

## Muscle activity

The nervous system and the muscles of the body can be investigated in a similar way to that described for the ECG. Muscle is made of motor units which consist of a single branching neuron from the brain or spinal cord and between 25 and 2000 muscle fibres (cells). The record of the electrical signals during movement is called the **electromyogram**, (EMG). Individual fibres cannot be investigated because of their small size, but a single motor unit can be measured by means of a needle electrode inserted into the muscle, or a small area can be measured with a surface electrode (Figure 5.11). The signals are produced by the action potentials being generated asynchronously, that is not in a connected way, like the heart. The EMG trace is therefore full of random spikes (Figure 5.11(d)), without the regularity seen in the ECG. Muscles can normally function for long periods without fatigue, provided that the frequency of restimulation allows a relaxation period of about 0.2 s. In the muscular disease *myasthenia gravis* this continued action is lost. The EMG can also be used to measure the speed of conduction of the action potentials in the motor and sensory nerves. A reduction in this, significantly below normal, indicates nerve damage. EMG signals can be used to control artificial limbs. The Utah arm, for example, uses signals from the shoulder muscles to control the opening and closing of an artificial hand.

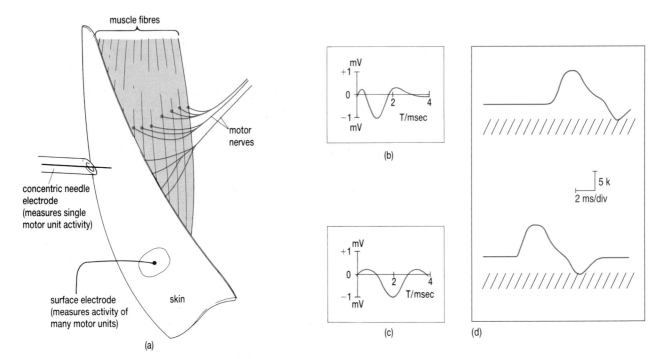

**FIG 5.11** **(a)** Measuring muscle activity **(b)** needle electrode electromyogram **(c)** surface electrode electromyogram **(d)** EMG trace.

## Brain activity

When you think of the concentration of neurons in the cortex of the brain, you will realise that the problem is to discern what is actually being measured, when electrodes are attached! These are usually small discs of chlorided silver in locations that, as with ECGs, are internationally agreed and labelled. The reference or neutral electrode is attached to the ear. The resulting **electroencephalogram** (EEG), is a trace which has a very low amplitude (about 50 μV), and a frequency which seems to depend on the mental activity of the subject. The small size of the signal can cause problems with noise from, for example, the muscle potentials from eye movements. As the brain frequencies are low the signal is usually passed through a filter to remove the higher frequency EMG signals, as well as being amplified for display or recording.

The signals are thought to originate from groups of neurons whose activity is normally synchronised to produce regular bursts of charge from different sites over a period of time. A waveform contains variations of both amplitude and frequency, but certain ranges or 'rhythms' are distinguishable. The common frequency range for a relaxed or lightly sleeping subject is 8 to 13 Hz, the alpha brain wave. When a person is more alert the frequency increases to the beta wave range, above 13 Hz. In deep sleep the frequency falls to the theta (4 to 7 Hz) or delta (0.5 to 3.5 Hz) ranges, Figure 5.12 (a) and (b). Sometimes a sleeping subject shows a higher frequency pattern called paradoxical sleep or rapid eye movement (REM) sleep, because the eyes move during this period. REM sleep appears to be associated with dreaming.

EEGs are used in research on sleep and other investigations into normal brain function as well as for diagnosing brain disorders. One of the most distinctive is the high amplitude spikes which are associated with epilepsy (Figure 5.13). They can also be used in monitoring the condition of the patient during surgery to determine the degree of anaesthesia, for example.

Other uses of electrical potentials include the study of the retina and the movements of the eye.

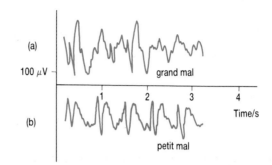

(a)

100 μV

early sleep (alpha wave)

1 2 3 4 5 6 7 8 9 10

Time/s

deep sleep (delta wave)

(b)

**FIG 5.12** EEG traces for stages of sleep **(a)** early sleep (alpha wave) **(b)** deep sleep (delta wave).

(a)

100 μV

grand mal

1 2 3 4

Time/s

(b)

petit mal

**FIG 5.13** EEG traces for epileptic attacks **(a)** grand mal **(b)** petit mal.

## 5.5 ELECTRICAL THERAPY

An indication of the currents which can be dangerous is given by school laboratory safety regulations which include the following recommendations: (i) in any work where there is a risk of the body coming into contact with mains supply, a circuit breaker should be fitted which will trip when 30 mA has flowed for 30 ms, (ii) equipment which generates voltages above mains should be protected by a trip or fuse which operates at a current of 5 mA.

As the body is operated naturally by biopotentials, it is not surprising that applied potentials have a wide range of uses to produce psychological, physical and sensory improvements in the body's functioning. In all of these cases the design of equipment must comply with stringent electrical safety regulations. In all cases there is the potential for harm as well as healing.

### Electroconvulsive therapy (ECT)

This is a treatment of the brain with electrical signals, under general anaesthetic and has produced many improvements in the behaviour of patients suffering from mental illness, particularly severe depression. The way it works is not well understood, and it may have undesirable side effects, such as short-term memory loss. Though widely used, it remains a controversial technique.

### Defibrillation

This is another dramatic intervention that is taken when the cardiac monitor shows that the heart is failing to pump the blood; the heart is in fibrillation. A current of about 20 A at 3000 V is passed through the heart for about 5 ms. This produces a major contraction of all the muscles in the heart, which usually jolts the organ back into its proper rhythm. The 'shock' is applied by placing two large electrodes, called paddles, above and below the heart. As with all of the uses of electrodes, it is important that there is good electrical contact with the skin. This is achieved with the use of a conductive paste. In this case it is also important that the electrodes are insulated from the operator!

**5.7**  (a)  How do you think the operator could be suitably insulated during defibrillation?

  (b)  How much energy (in joules) is passed through the heart in defibrillation?

  (c)  (i)  What is the resistance of the circuit through the body in this case?

    (ii)  Where is most of this resistance located?

  (d)  Draw a defibrillation circuit which operates from the mains supply and incorporates a capacitor across the electrodes to prevent sparking.

## Pacemakers

The pumping action of the ventricles is controlled by the atrioventricular node; if this is damaged pumping can go on to an autocontrol at a rate of about 30 beats per minute. At half the normal rate this is sufficient to maintain the life only of a very inactive person. To improve this situation the artificial pacemaker has been developed which produces a suitable electrical pulse 72 times a minute. It is placed in a cavity under the shoulder, by lifting a flap of skin, often under local anaesthetic. The wire to carry the signal is fed into a vein and advanced into the right ventricle by viewing the operation under X-ray fluoroscopy (details of fluoroscopic examination are given in Section 9.3). The pacemaker is powered with batteries that last several years and is enclosed in material that is impervious to body fluids. It has become a routine treatment in recent years.

## Functional electrical stimulation

This is a developing field of research showing some promising applications. Nerves or muscles are electrically stimulated to restore their function. A simple example is for people suffering from a condition called 'foot-drop'. The toes do not lift properly and the muscle responsible can be stimulated using surface electrodes. The user simply wears a belt which holds the power supply. More substantial applications may require implanted stimulators. Examples include those which enable people with major loss of lower body functions (paraplegics) to be able to stand, operate a tricycle for mobility or control their bladders (Figure 5.14).

**FIG 5.14** Stimulators for implantation.

Cochlear implants are a more established application, but one which is still being developed. Used to restore hearing in people whose cochlea is damaged, the implants stimulate the auditory nerve to produce the effects of sound. Details of how the ear works are given in section 3.1.

## Diathermy

This is the technique of local heating of tissue by a high frequency source. Two methods are currently used, *short wave diathermy*, which uses an electric field oscillating at frequencies of around 30 MHz, and *microwave diathermy* which uses electromagnetic radiation of frequency 2450 MHz. In short wave diathermy the tissue to be heated is placed either between the plates of a capacitor or close to an inductor. Thus electrical heating is produced either by the changing field in the capacitor or the eddy currents around the inductor. Microwave radiation is absorbed directly into the tissue, and the frequency can be chosen to interact with particular molecules – in this case water. This interaction is thermal (it makes the molecules move more violently), rather than chemical as in the ionising radiations described in Theme 3; however, care must be taken not to cook the exposed tissue! Diathermy is used in the treatment of joints (in arthritis), muscles and tendons (in strains and other traumas). Many sports teams have a portable diathermy kit available to apply emergency treatment on the field.

## Electrocautery and electrosurgery

More popularly known as burning and cutting (!) these common surgical tasks are often conveniently carried out with high frequency (>2 MHz), high voltage (~15 kV) electricity. This is supplied to a fine probe which is then used to stop bleeding during surgery by coagulating blood in vessels that are too small to tie (electrocautery), or to cut through tissue (electrosurgery).

Care must be taken that there is no burning at the other electrode (which connects the patient to the supply) and no shock hazard. The power density at the tip can be as high as 3300 W cm$^{-3}$. The cutting is produced by the extreme local heating (up to 800 °C) which results in the physical rupturing of tissue by the rapid boiling of fluids. If the probe is moved rapidly the destruction can be limited to a depth of 1 mm from the probe. Electrosurgery is used in operations on the brain, bladder, prostate, spleen and cervix, often in conjunction with an endoscope (see Chapter 7).

### SUMMARY ASSIGNMENTS

**5.8** Draw up a table to compare the values of the electrical quantities mentioned in all sections of this chapter. These would include potential difference, current, frequency, power, time period, energy and perhaps others. Not all this data is given but you should be able to work some of it out. What conclusions (if any!) can you draw from these figures?

**5.9**

**FIG 5.15**

The graph Figure 5.15 shows a normal electrocardiogram (ECG) signal obtained at the surface of the skin of a patient.
**(a)** What is the amplitude of the main pulse in the signal?
**(b)** Find the period of the heart beat and from it calculate the pulse rate per minute.
**(c)** What changes would you expect to see in the electrocardiogram if the patient began to take exercise?
**(d)** Sketch the graph and label with a P the points where atrial contraction occurs and with a Q the points where ventricular contraction starts.

*(AQA Specimen 2001/2)*

**5.10 (a)** The potential inside the membrane of an axon is known as the membrane potential. The curve on the graph shows how the membrane potential varies with time when the nerve cell to which the axon is connected is stimulated.

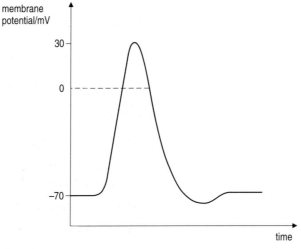

**FIG 5.16**

Sketch the graph and
**(i)** add a suitable time scale for this process,
**(ii)** label the part of the curve which shows depolarisation,
**(iii)** label the region where the membrane potential has a reversed polarisation.

Write down
**(iv)** the value of the resting potential
**(v)** the ions which diffuse into the axon during depolarisation
**(vi)** the ions which diffuse out of the axon during repolarisation

**(b)** Explain briefly how the action potential generated by the process described above travels along a nerve fibre carrying a signal to or from the brain.

*(NEAB June 1998)*

# 6 PRESSURE

iving at the bottom of an ocean of air many kilometres deep, we are subjected to the considerable pressure of the atmosphere. All parts of the body have to operate under this pressure. Fluid systems that require pressure to drive them must have pressure differences relative to this. When we breathe in or suck a drink through a straw, the pressure in the body will be less than atmospheric, the heart pumps blood at greater than atmospheric pressure. In this chapter we will be continuing the emphasis on the circulatory system, whose condition can be monitored in some detail by pressure measurements.

## LEARNING OBJECTIVES

*After studying this chapter you should be able to:*

① give a simple description of the human circulatory system, including typical blood pressures and flow rates, and give examples of the diagnostic importance of their measurement;

② describe the use of the sphygmomanometer to measure blood pressure;

③ give examples of invasive methods of measuring blood pressure, including those using resistance, capacitance and inductance transducers;

④ explain the physical principles of resistance, capacitance and inductance transducers;

⑤ give examples of other fluid pressures within the body.

## 6.1 MEASURING PRESSURE

n biomedical measurement, pressures are usually given exclusive of atmospheric pressure. This is called the gauge pressure in contrast to the absolute pressure. Thus:

gauge pressure = absolute pressure – atmospheric pressure.

# Units

Pressure is defined as the force per unit area, so its unit in SI is N m$^2$, which is given the name **pascal (Pa)**. This is unfortunately a rather small unit; atmospheric pressure for example is about 10$^5$ Pa, so you will often find the unit kPa (kilopascal) used. 10$^5$ Pa is called a bar.

In a liquid or a gas the pressure can be calculated from:

pressure ($p$) = density ($\rho$) × depth ($h$) × gravitational field strength ($g$)

(For a derivation of this formula see a physics course text.)

The basic method of measuring pressures in the body is using a manometer. You may have measured your lung pressure by blowing into a water U-tube manometer and reading off the height of the column. Thus the pressure is given in centimetres of water (cm H$_2$O), or alternatively in millimetres of mercury (mm Hg) because it has a particularly high density (though it is a health hazard to carry out this direct measurement with mercury in the manometer). The direct measurement of blood pressure is rather more difficult, involving the insertion of a catheter tube into the blood vessel, which is called an *invasive* technique. The Bourdon pressure gauge (Figure 6.1), which is like the aneroid barometer, is now usually used, rather than a manometer. Normally however, a non-invasive technique is used, involving the sphygmomanometer. This is described in the next section.

Direct measurement of the circulation system was first carried out by Rev. Stephen Hales in London, in 1733. Using a sharpened goose quill to pierce with, and a goose trachea to make a flexible connection, he connected a 9 foot vertical glass tube to the artery of a horse, and found that the blood rose to an average height of 8 feet above the heart.

**FIG 6.1** Bourdon gauge for measuring pressure.

**TABLE 6.1** Normal body pressures – typical values

| | | TYPICAL PRESSURE /mm Hg |
|---|---|---|
| Arterial blood pressure | – Maximum (systole) | 100–140 |
| | – Minimum (diastole) | 60–90 |
| Venous blood pressure | | 3–7 |
| Great veins | | <1 |
| Capillary blood pressure | – Arterial end | 20–35 |
| | – Venous end | 10–20 |
| Middle ear pressure | | <1 |
| Eye pressure – aqueous humour | | ~20 |
| Cerebrospinal fluid pressure in brain (lying down) | | 5–12 |
| Gastrointestinal pressure | | 10–20 |
| Intrathoracic pressure (between lung and chest wall) | | –10 |
| The atmosphere | | 760 |
| or  1.01 × 10$^5$ Nm$^{-2}$,  1.01 × 10$^5$ Pa,  1.01 bar  1033 cm H$_2$O,  14.7 psi,  1 atm | | |

## Values

Table 6.1 gives typical values of pressures within the body. You will note the highest fluid pressure is at the output of the heart, and that the pressure falls in stages through the circulation system.

The *cerebrospinal fluid* (CSF) cushions the brain within the skull. In infants, there is sometimes a blockage in the outlet for the fluid to the spinal cord. The excess fluid can cause the condition of *hydrocephalus* ('water on the brain'). It can be detected by the increased pressure it causes and corrected with an operation to install a bypass drainage system. CSF pressure is not usually measured directly.

The fluids in the eye are under pressure to keep the eyeball in a fixed size and shape. In Chapter 2 it was shown how the clarity of vision is very dependent on this. When you press on your eyelid gently you can feel this pressure. If you partially open your eye as you do this you may notice a loss of focus. Pressure build up in the eye can cause a condition called *glaucoma*, which results in tunnel vision and in some cases blindness. The pressure is usually measured with an instrument called a tonometer, which measures the indentation of the eye by a known force.

The pressures in the gastrointestinal tract can vary considerably as a result of the contents of the various parts and the action of various valves, which are designed to permit food to flow in only one direction. Blockages of these valves such as the pylorus between the stomach and the gut, or of other parts of the gut will increase the pressure and this can have fatal consequences so surgery may be necessary. The pressure in the digestive system is linked to that of the lungs through the flexible diaphragm that separates the two organ systems. Thus the pressure of the digestive system can be increased voluntarily by taking a deep breath, closing the windpipe at the glottis, and contracting the abdomen (as in defecation).

The highest pressures in the body are to be found in the load-bearing bone joints as described in Chapter 1, and the pressure would be even higher if the joints were not as large as they are. The frequent damage to cartilage suffered by people in sporting activities is a measure of the exacting conditions under which it lubricates the joints.

---

**QUESTION**

**6.1**   Calculate the values of some of the body fluid pressures listed in Table 6.1, in kPa.

---

## BLOOD PRESSURE

### Blood flow

The circulatory system has been described in Section 5.2, and illustrated in Figure 5.5 which shows how the pressure changes around the body. There is a corresponding change in the blood flow as the branching vessels take the blood into ever smaller arteries and then into veins of increasing size. A summary of the dimensions, blood flow velocities and blood pressures is given in Table 6.2. The values are averages, for easy comparison. You will notice that veins are wider than arteries because the blood flows at a much slower speed, yet the total flow rate of the circulation must of course be constant. The veins hold about 80 per cent of the volume of the blood (about 5 litres for an adult) at any one time, so they serve as a kind of reservoir. The highest flow is at the aorta and pulmonary artery, where between 3.5 and 5 litres per minute flow in a normal adult at rest. This is called the *cardiac output*. In the capillaries by contrast the flow is so slow that the movement of individual blood cells can be observed under the microscope.

There is a direct relationship between pressure, flow rate and resistance in the blood vessels that is analogous to the electrical relationship. However, there are other factors that affect flow. The small blood vessels near the body surface, the pre-capillary arterioles, can be narrowed (*vasoconstriction*), or widened (*vasodilation*), to respond to body temperature control, as described in Chapter 4. Certain drugs such as nicotine can also produce vasoconstriction.

**TABLE 6.2** Cardiovascular system – typical values

| VESSEL | NUMBER /1000 | DIAMETER /mm | LENGTH /mm | MEAN VELOCITY /mm s$^{-1}$ | PRESSURE /mm Hg |
|---|---|---|---|---|---|
| Aorta | – | 10.50 | 400 | 400 | 60–140 |
| Terminal arteries | 1.8 | 0.60 | 10 | <100 | 35–50 |
| Arterioles | 40 000 | 0.02 | 2 | 5 | 40–25 |
| Capillaries | >million | 0.008 | 1 | <1 | 25–12 |
| Venules | 80 000 | 0.03 | 2 | <3 | 12–8 |
| Terminal veins | 1.8 | 1.50 | 100 | 1 | <8 |
| Vena cava | – | 12.50 | 400 | 200 | 3–2 |

The flow is not constant throughout the cross-section of a vessel, but increases as the distance from the wall increases. Detailed information of the flow profile can be obtained by Doppler ultrasound, which is described in Chapter 8. It can indicate the condition of the vessel walls. Obstructions can build up from fatty deposits on the walls (*atheroma*), when these form a complete blockage (clot or *thrombus*), the flow of blood to a part of the body is prevented. If this is in the brain the result is a *stroke*, in the heart muscle it is called a *coronary* or a heart attack. Clearly these are potentially fatal conditions, so the measurement of blood flow is very important.

## *Pulse*

The pumping action of the heart causes a regular pulsation in the blood flow which can be felt in an artery near the surface, such as at the wrist. This is the pulse, and a typical rate is about 70 per minute. We will now consider in more detail how the pressure changes with time. As the ventricles contract, they increase the pressure in the arteries leaving the heart, until this back pressure closes the valves from the heart. The blood then runs away through the arteries, so the pressure falls in the aorta and pulmonary artery. A pressure wave then passes around the circulatory system, getting smaller as it progresses to bring the venous blood back to the heart. The cycle then begins again, with the rising pressure period of about one third of a second, which is called the *systole*, being followed by the longer falling pressure period, the *diastole*. Figure 6.2 shows how this cycle relates to the electrical signals of the ECG. These changes are dependent on the physical condition of the heart and the blood vessels, so detailed measurement of blood pressure is a valuable diagnostic tool. For example, high blood pressure (*hypertension*), is a condition that can lead to heart failure, if it is not treated.

**FIG 6.2** Pressure variation in the aorta, during one heartbeat, and associated ECG.

---

**QUESTION**

**6.2** We can calculate the rate of working (the power) of the heart, from a knowledge of the flow rate and pressure of the blood.

Suppose blood flows with pressure $p$ with flow rate $v$ per second, through a major artery of area $A$ (Fig 6.3):

**FIG 6.3** Schematic diagram of blood flow.

(a) The work done by the blood per unit time is force ($F$) multiplied by distance travelled per unit time ($l/t$). Write down expressions for $F$ and $l/t$ in terms of $p$, $A$ and $v$.

(b) Show that the power $P = p \times v$

(c) Calculate the power developed in an artery which has blood flowing at a rate of 5 dm$^3$ min$^{-1}$ and a pressure of 125 mm Hg.

---

## Sphygmomanometer

This instrument is used for routine blood pressure measurements. It is easy to use and involves no direct contact with the bloodstream. It is however less accurate than those methods that do, is not suitable for continuous monitoring, and gives only the systolic and diastolic pressures, with no indication of the shape of the waveform shown in Figure 6.2. The method often fails when the pressure is very low, for example, when the patient is in a state of shock.

The instrument consists of an inflatable cuff and a manometer. The cuff is strapped around the arm at heart height. Air is pumped in manually until a pressure of 200 mm Hg is applied to the brachial artery, so that it obstructs the blood flow. When a stethoscope is placed on the artery below this point no sound is heard (Figure 6.4). The pressure is slowly reduced by releasing air until the blood begins to spurt through a small opening in the artery, when it is at its maximum pressure. This causes turbulence and the sound, called *Korotkoff sound*, can be heard in the stethoscope. This highest, systolic pressure can then be read on the manometer. These sounds continue as the pressure falls, until the blood vessel remains open throughout the pulse period. This means that the applied pressure has now fallen as low as the lowest, diastolic pressure. So a second manometer reading is taken when the Korotkoff sounds cease. The technique of listening for the sounds is called *auscultation*.

**FIG 6.4** A doctor measuring a patient's blood pressure with a sphygmomanometer.

Electrical or automatic sphygmomanometers have been devised which incorporate a microphone or a pressure transducer to detect the onset and end of the blood turbulence as the pressure is reduced.

## INVESTIGATION

### Measurement of blood pressure

Use a manual or electronic sphygmomanometer to measure blood pressure. Try to identify factors which increase and decrease blood pressure. Things to try include exercise and caffeine to increase heart rate, and deep breathing or meditation to decrease.

## QUESTION

**6.3**   (a)   How reliable is the sphygmomanometer? What do you think are the major sources of inaccuracy? How would an automatic system remove these?
   (b)   Why should the operator check that the position of measurement is at the same height above the ground as the heart?

**B**ecause of the shortcomings of the indirect methods it is often necessary to introduce a transducer directly into the bloodstream to measure pressure. This is done by inserting a **catheter** into the appropriate vessel *percutaneously*, that is through the skin. The catheter is a plastic tube which can be passed via the superficial (near surface) veins or arteries, throughout the circulatory system and into the heart. The external end is connected to a pressure transducer (Figure 6.5). When a rapid frequency response is required the transducer can either be mounted at the internal the end of the catheter or introduced while it is in place.

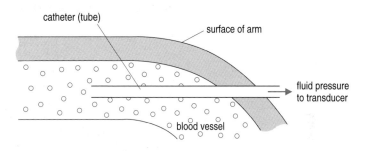

**FIG 6.5** Internal measurement of blood pressure.

## Transducer types

The principle of the pressure transducer is that it produces a change in an electrical circuit as a result of deflection caused by the pressure of the blood. A thin, flexible **diaphragm** (of stainless steel, or phosphor bronze) has the transducer attached to it. There are a number of different types available. The selection of instrument will depend on the measurement site and the size and range of the pressures to be measured. For example the pressure in the arteries is high (up to 120 mm Hg) and changes rapidly, so a high frequency response is important. In the veins the fluctuations are slower and the pressures much lower (from 0–10 mm Hg), so the sensitivity of the instrument is more important. If the transducer is placed in the bloodstream it must be coated with plastic to prevent clotting. The properties of resistance, capacitance and inductance are all used in pressure transducers. The resistance transducer is the most common, and is often called a **strain gauge**.

## Resistance transducers

When the dimensions of a wire change because of stretching, its resistance shows a corresponding change, since

$$R = \rho L/A$$

The resistance elements are attached to a deflectable diaphragm, to produce a distortion or strain. The devices are compressed or extended, Figure 6.6(a) to follow the strain of the diaphragm hence the name strain gauge.

The transducers may be of the **bonded** or **unbonded** type, (Figure 6.6). In the bonded type the wire is arranged in a zig-zag pattern and firmly bonded to the diaphragm so that it moves in the same way. In the unbonded type one end of the wires are connected to the diaphragm and the other to a membrane or frame which can move relative to the diaphragm. This causes two of the wires to stretch and the other pair to relax. With very thin wires this can provide a sensitive response.

In order to compensate for temperature fluctuations, the gauges are usually arranged in a bridge circuit, Figure 6.6(c). The four have the same resistance when unstrained and will be affected equally by any temperature changes. When the diaphragm moves, the change $\Delta R$ is then the result only of the dimension change. The transducer inside the body is connected to the bridge circuit and its power supply, and can be calibrated to read pressure directly.

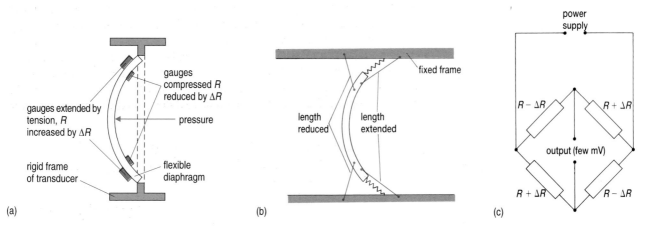

**FIG 6.6** Measurement of strain gauge resistance changes **(a)** position of gauges on diaphragm (bonded type) **(b)** unbonded strain gauge **(c)** bridge circuit.

The alternative is to use a semiconductor. The silicon bonded gauge has been widely used for its sensitivity to very small displacements. For example with a supply voltage of 10 V it can produce a 30 mV signal from a 3 μm displacement, over a pressure range of 300 mm Hg. The sensitivity of a pressure transducer is usually expressed as microvolts per (applied) volt per mm Hg. Alternatively the **gauge factor** is used to express sensitivity, this is the ratio of the fractional change in the resistance to the fractional change in the length:

$$G = \frac{\Delta R/R}{\Delta L/L}$$

The gauge factor of a metal is typically 2, that of a semiconductor about 150. Semiconductor gauges are very temperature sensitive however, so they require a compensation circuit where temperature fluctuations are large.

---

### QUESTION

**6.4** **(a)** What is the pressure sensitivity of the silicon bonded gauge quoted here?

**(b)** A semiconductor pressure transducer of length 3 mm is extended by 3 μm in use. What would be the extension of a metal wire transducer of length 8mm, to produce the same fractional change in the resistance? (Use the gauge factors given above.)

---

## Capacitance transducers

The capacitance of a parallel plate capacitor is given by:

$$C = \varepsilon A/d$$

Where $d$ is the distance between the plates, $A$ is the area of overlap and $\varepsilon$ is the permittivity of the insulator between them, called the dielectric. If any of these is changed, by attaching one plate or the dielectric to the diaphragm, (Figure 6.7) then the capacitance will change. This is the principle of the capacitance transducer. The diaphragm itself may form one plate of the capacitor, separated from the fixed plate by a few hundredths of a millimetre. The capacitance is measured using an a.c. circuit. The transducer may form one arm of a bridge circuit so that the imbalance gives a measure of the capacitance. Alternatively, included in a tuned circuit it will vary the resonant frequency of the circuit. A third method is to measure the voltage across the capacitance transducer when it is in series with an inductor. Compensation for temperature changes is achieved in the same way as with resistance transducers, by using identical capacitors on opposite arms of an a.c. bridge, so that one increases its capacitance whilst the other reduces, when the pressure is applied. This arrangement is called a differential capacitance transducer.

**FIG 6.7** Capacitance transducers **(a)** variable separation **(b)** variable area **(c)** variable permittivity.

The advantage of this type of transducer is that it can have a very linear shape and so not obstruct the flow of blood. The displacement is very small so they are very sensitive and have rapid response times. They are however more difficult to use than the resistance type and suffer from interference from the capacitance of the lead wires.

**FIG 6.8** A linear variable differential transformer.

## Inductance transducers

The inductance of a coil depends on the amount of permeable core within it. If this core is moved then the self-inductance of the coil will vary. Thus the inductance will depend on the pressure which moves a diaphragm connected to the core. Changes in inductance are measured by an a.c. bridge, a series circuit or an oscillator circuit. The core can link a primary and a secondary coil, so that the mutual inductance changes, but the most common arrangement is to use two identical secondary coils as shown in Figure 6.8. This is called a linear variable differential transformer (LVDT). The primary coil is given an audio frequency signal which induces voltages in the two secondary coils in opposition to each other, as they are wound in series and in opposition. All three coils are linked by the core of a high permeability ferromagnet, which is attached to the diaphragm. As the core moves from the balanced null position, the voltage induced in one coil increases whilst the other decreases. This results in an output voltage which varies linearly with the position of the core and therefore the pressure. The core can be spring-loaded to accept pressure from one side, or it can accept pressure from both sides and measure the pressure difference between two points.

The LVDT is sensitive and has a rapid response. It is also insensitive to temperature changes and quite robust. It is however rather larger than the two preceding types.

## Piezoelectric transducers

These are made of materials which generate a potential difference when they are subjected to pressure. They are active transducers, requiring no external power supply, as noted in Chapter 4, in contrast to the transducers previously described. However they cannot measure constant pressure, only respond to changes in pressure. They have an important role in the measurement of blood flow rate in Doppler ultrasound. This is described in Chapter 8.

## SUMMARY ASSIGNMENTS

**6.5** Explain which of the following types of transducer are, in their basic form, capable of measuring displacement and which measure pressure by measuring velocity: piezoelectric crystal, LVDT, capacitance transducer, bonded strain gauge, unbonded strain gauge.

**6.6** List the factors which are important in choosing suitable methods of measuring blood pressure. What are the advantages of the methods described in this chapter, in relation to these factors?

**6.7** The diagram Figure 6.9 shows an incomplete sphygmomanometer which is a device used for blood pressure measurement.

**FIG 6.9**

**(a)** Three valves are required to allow the balloon to be inflated and deflated. On a sketch of Figure 6.9, mark the positions of the three valves at suitable places and label them $V_1$, $V_2$ and $V_3$. State what each valve is used for.

**(b)** If the air pressure is taken as 760 mm of mercury, what is the pressure, in mm of mercury, at points A, B and C on the diagram?

**(c)** Describe the procedure for obtaining the systolic blood pressure reading using this instrument.

**(d)** A systolic blood pressure reading for a patient is given as 120 mm of mercury. What is this value expressed in pascal? (density of mercury = $1.36 \times 10^4$ kg m$^{-3}$.)

**(e)** This blood pressure reading was, as usual, measured in the arm at the same level as the heart. Estimate the pressure, in pascal, in the foot of this adult patient if she was standing. (density of blood = $1.06 \times 10^3$ kg m$^{-3}$.)

**(f)** Fainting is caused by reduced blood flow to the brain. Explain why one suggested treatment is to lower the patient's head.

*(NEAB June 1997)*

**6.8** The diagram Figure 6.10 shows the flexible diaphragm of a pressure transducer designed to measure blood pressure variations. Four resistance gauges $R_1$, $R_2$, $R_3$ and $R_4$ are bonded to the diaphragm so that they change in length as the diaphragm flexes. When there is no pressure difference between the surfaces of the diaphragm, it is flat.

**FIG 6.10**

Blood pressure is applied from the left as shown, causing the diaphragm to bend.

**(i)** Which resistance gauges will be increased in length and which decreased?

**(ii)** When the diaphragm is not bent each resistor has a resistance of 10 $\Omega$. If the lengths of the resistance gauges are changed by 5% when the diaphragm is bent as shown, calculate the new resistance of each gauge. Assume there is no change in cross-sectional area of the resistors.

**(iii)** To register the pressure change, the four resistance gauges are connected in a circuit which converts the changes in resistance into a voltage change. Without giving details of a suitable circuit, explain why it is necessary to use more than one resistance gauge to obtain a reliable pressure measurement.

# 7

# OPTICS

The investigation of the human condition using light can hardly be avoided. All patients are seen by their doctors and external appearances are used in many diagnoses. Internal observation could obviously bring much additional information. Few opportunities are as simple as the mouth, all that is required here is to 'open wide and say "aah"'. Specially focused beams enable internal viewing of the eye (with an ophthalmoscope) and the ear (with an auroscope). In most cases a tube (an endoscope) has to be inserted into an orifice, to see inside the body.

Good health has traditionally been associated with exposure to light. During the industrial revolution in Britain in the nineteenth century, air pollution and working conditions reduced this to the extent that many people suffered diseases caused by lack of vitamin D. This is manufactured by the body by absorbing sunlight. Now there is more concern about the dangers of too much sunlight. An increase in the incidence of skin cancer has been associated with the greater exposure of people with light coloured skin to strong sunlight – for example on beach holidays. This is an example of the ionising properties of the ultraviolet radiation in sunlight, details of which are given in Theme 3.

Laser light provides the opportunity to expose the body to light in a very specific way (Figure 7.1). The intense beam is able to cut (incise), and burn (cauterise), tissue in a way which makes the laser a useful addition to the surgeon's toolkit.

Ultraviolet radiation (UVR) is strongly linked to the induction of skin cancers and probably to eye disorders, in particular cataract, and the suppression of the body's immune system. We live in an environment where UVR is usually present, the main source of exposure coming from the sun. Scientific studies cannot demonstrate that there is a completely safe level of UVR.

*National Radiological Protection Board*

## LEARNING OBJECTIVES

*After studying this chapter you should be able to:*

1. use the following scientific terms correctly: total internal reflection, coherent, numerical aperture, resolution, laser;

2. explain the principles of the transmission of light by optical fibres;

3. describe how the fibre optic endoscope is constructed; compare the operation of the video endoscope which uses a charge coupled device;

4. give examples of the use of endoscopes in diagnosis and treatment;

5. explain the principles of operation of the laser and give examples of its medical application.

**FIG 7.1** An argon laser removing a birthmark.

## FIBRE OPTICS

## *Transmission*

When light enters the end of a glass rod, it will be **totally internally reflected** if it strikes the surface of the rod at greater than the critical angle (Figure 7.2). These reflections can continue along the length of the rod so that the light emerges with little loss of intensity. If the glass is made into thin filaments it is surprisingly flexible, and can carry light to inaccessible places. A bundle of these filaments produces the **fibre optic light guide**.

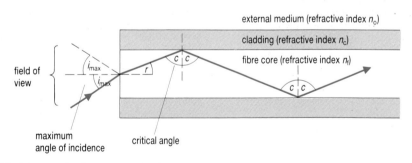

**FIG 7.2** Internal reflection in a glass fibre.

The refractive index $n_m$ of a medium is given by

$$n_m = \frac{\sin i}{\sin r}$$

where $i$ is the angle of incidence of a ray of light in air or a vacuum, on the surface of the medium, and $r$ is the angle of refraction within the medium.

Light can easily be lost from the surface of the fibre if it is in contact with a substance of equal or higher refractive index. This could be grease or an adjacent fibre. To avoid such leakage the fibres are clad with glass of a lower refractive index than the fibre core. The selection of the core and cladding materials affects the performance of the fibre. In particular, the maximum angle of incidence from which the light can be admitted, $i_{max}$, is affected. $i_{max}$ is called the half angle of the fibre since it represents half the field of view that can be transmitted. With reference to Figure 7.2 it can be shown that

$$n_0 \sin i_{max} = \sqrt{(n_f^2 - n_c^2)}$$

The term $n_0 \sin i_{max}$ is called the **numerical aperture**; a typical value is 0.55, giving a half angle $i_{max}$ of about 33° in air. It can be seen that the numerical aperture increases with an increase in the difference between the refractive indices of core and cladding. There is a limit to the difference that can be produced however, since the absorption of light increases if the refractive index of the core is above a certain value.

This analysis is strictly true only when the fibre is straight, any curve will of course change the angle at which light strikes the surface. In practice curves as sharp as about twenty times the fibre diameter result in only minor losses of light.

Intensity losses are usually quoted in dB m⁻¹ (see Chapter 3 for details of decibel scale). Typical values are:
ordinary window glass: 0.1 dB m⁻¹
best optical fibre: $2 \times 10^{-4}$ dB m⁻¹.

## QUESTION

**7.1**    An optical fibre has a refractive index of 1.55 and is clad in glass of refractive index 1.42. Calculate:
   **(a)** the numerical aperture of the fibre,
   **(b)** the half angle of the fibre in air.

**FIG 7.3** A coherent bundle of fibres.

**FIG 7.4** Production of a fibre bundle.

## Image formation

If the bundle is to be used to transmit an image of the inside of the body, then attention has to be paid to a number of factors. There will be a loss of intensity due to absorption and scattering by the glass, which will depend on the length of the fibre. Production methods are now such that there is very little loss over the length required for observing parts of the body (maximum length about 1 m).

There will also be losses at the entry to and exit from the fibre, and at the reflections along the length of the fibre.

Each fibre will transmit the light incident on its face, independent of all the other fibres. To transmit an image the fibres have to be kept in the same relative positions, however the bundle is bent (Figure 7.3), so that each fibre's element makes the correct contribution to the whole image. This is called a coherent bundle. The resolution of the image, the amount of detail that can be seen, depends on the fineness of an individual fibre, and how closely they are packed together. There is a limit to this; if the fibre diameter is very small, diffraction effects interfere with the linear transmission. A typical diameter is 10 μm with about 60 per cent of the total area of the cross-section of the bundle occupied by core (this is called the packing fraction). The size of the bundle depends on its use, some examples are given in the next section. Diameters range between 0.5 mm and 3 mm, containing between 5000 and 40 000 fibres.

In a bundle designed to transmit light rather than an image, the fibres can be larger and incoherently packed. The fibres are typically 30 μm in diameter, which makes them more efficient at transmitting light. It also makes the light guide much cheaper to produce than the coherent image guide.

The manufacture of the fibre bundle is an automated process. An individual glass rod is coated with the low refractive index glass and bundled together with thousands of others in a cover sheath of glass. This large bundle is heated and drawn to reduce the size. It takes many passes until the rods are reduced to the fibre diameter of about 10 μm (Figure 7.4). The glass cover has fused during this process and kept the fibres coherently arranged. It is then dissolved away in acid to leave the fibres free to be flexible, but fixed in position by an external sheath.

In the Figure 7.4 labels: single rods, cover sheath, electric furnace, fibre conduit treatments. Examples include: using forceps for taking samples of tissues (called biopsies), electrodes to apply heat to stop bleeding and various snares or extractor devices to remove obstructions such as foreign bodies (Figure 7.8).

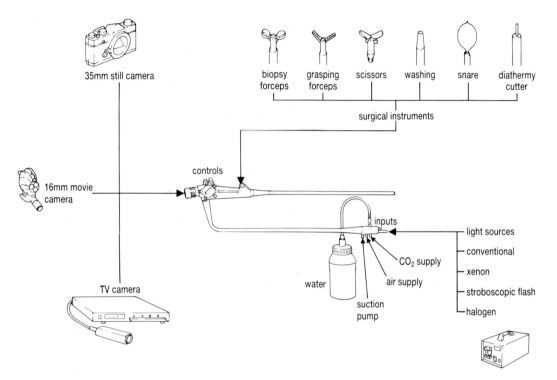

**FIG 7.7** Some accessories for an endoscope.

**FIG 7.8** Endoscope image showing the removal of the top of a martini stirrer from the windpipe of a young child who had accidentally swallowed it.

### QUESTION

**7.3** Write a brief description of some of the uses of endoscopes for diagnosis and treatment, with reference to this passage and illustrations.

## *The video endoscope*

In America in 1983 a radically different endoscope was shown in action for the first time. It makes use of a device which converts the received light into an electrical signal at the distal end of the scope. This is the **charge coupled device (CCD)**, originally developed to measure low levels of light detected in astronomy. The principle is like measuring the distribution of rainfall in a field by setting out an array of buckets before the rain and afterwards moving them on conveyor belts to a metering station where the amount in each bucket is recorded. The buckets are equivalent to small zones of low potential below an array of electrodes formed on the surface of a thin wafer of silicon. When the photons strike the silicon they produce charge in each potential well in proportion to the number of photons, (Figure 7.9). Each well becomes a small electrical image or pixel. The images are read off by sequential changes in the voltage on the electrodes placed across the grid. This is the charge coupling, and it enables the information from individual small areas to be recorded and used to build up the complete image. This electrical signal is passed by wire to a video processor which digitises the information for display on the TV monitor and for video or computer tape storage. The endoscope tube is similar to the fibre optic tube, except for the absence of the image guide bundle. The current designs of CCD have between 32 000 and 56 000 pixels, which is twice the number of fibres in a typical fibre endoscope. Use can also be made of the wavelength sensitivity of silicon, to identify particular types of tissue by the use of special photoactive compounds. The main advantage of the system is that of a permanent display and storage of everything illuminated. The major disadvantage is, as always with new technology, the price, but in time this is certain to reduce.

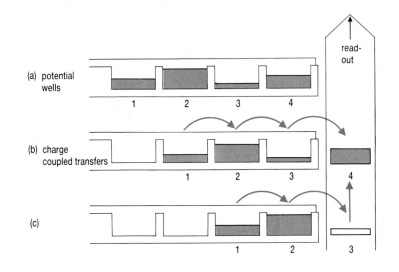

**FIG 7.9** A schematic diagram of the action of a charge coupled device.

> **QUESTION**
>
> **7.4**   Explain in your own words, the advantages and disadvantages of a video endoscope system. How do you think this will change in say the next ten years?

## 7.3   LASERS

The laser offers the opportunity to deliver energy to the body at a high intensity, and a very precise location, usually with an optical fibre, and internally by use of an endoscope. The energy can be used to cut, coagulate or destroy tissue, so that the laser can be used as a surgical tool. A further advantage is that lasers produce light of a particular wavelength. The interaction between tissue and light is very dependent on the wavelength of the light, the total energy absorbed, the rate of energy supplied, and the type of tissue. Over the past thirty years, this has resulted in the development of a range of lasers for different medical applications.

## Laser – the coherent substitute for a scalpel?

Laser is an acronym for Light Amplification by the Stimulated Emission of Radiation. Stimulated emission depends on changes in the energy of electrons orbiting the nuclei of the atoms of the lasing medium.

By absorbing a photon, an electron can move to a more energetic orbit around the atom. This 'excited' state is less stable and tends to lose energy by spontaneous emission of a photon of light. The energy of the photon emitted, and hence its wavelength, depends exactly on the difference in energy between the electron's orbit. If a photon with this amount of energy hits an excited atom, it is absorbed and re-emitted, together with a second photon of exactly the same wavelength and phase as the first. This phenomenon is stimulated emission. The emitted photons travel in the same direction as and in phase with the incoming photon. (Fig 7.10)

**FIG 7.10** The lasing process.

In a laser, the aim is to ensure that more atoms in the lasing medium are in an excited state than in the lower-energy state. To this end, light or electrical energy is pumped into the lasing medium. When excited atoms are in the majority, a population inversion is said to exist. A stray photon of the correct wavelength, produced by spontaneous emission, is enough to set off a chain of stimulated emissions (hence light amplification). Because the lasing medium lies between two mirrors, one totally reflecting and one partially reflecting, photons can bounce back and forth, stimulating more and more atoms to emit photons, thus rapidly increasing the intensity, as they leave through the partially reflecting mirror. If the pumping energy is applied continuously, population inversion is maintained and gives rise to a continuous-wave laser. If the pumping energy is applied intermittently, as in a pulsed laser, the stimulated emissions die down as the atoms that are in an excited state are depleted, destroying the population inversion.

It is possible to reduce the laser cavity's ability to bounce photons back and forth during pumping by inserting a shutter between the mirrors. This technique termed spoiling the quality (Q) of the oscillation of photons, allows a very high population inversion to build up. When the shutter is open, all the energy leaves in a short pulse of very high power that lasts about 10 nanoseconds. Such a device is called an 'A-switched laser'; it is possible to A-switch most pulsed lasers by fitting a shutter inside the laser cavity.

The average power output of a continuous-wave laser and of a pulsed laser is about the same, but the power of the pulsed laser at the peak of its pulse is much greater than that of the continuous model. (Turning a laser on and off very quickly does not produce the same effect as a pulsed laser.)

The arrangement of the mirrors at either end of a laser cavity ensures that the light produced by the laser is collimated (parallel) and coherent (the waves are in phase). Photons emitted in other directions are lost through the walls of the laser cavity. The light produced is monochromatic (of one wavelength), as it is composed of many photons that have exactly the same energy. The colour depends on the nature of the lasing medium.

These properties suit the laser to medical purposes very well. Because the beam is collimated, it can pass easily along optical fibres within an endoscope, a narrow tube that a doctor can pass into the lungs or the alimentary canal, for example. As the beam is monochromatic, it will be absorbed by some kinds of tissues and not others. This property is useful in surgery on the eye, for example, when the blue-green beam of the argon-ion laser passes without effect through the cornea and lens in order to treat the retina.

The name of laser refers to the lasing medium used. Lasers that rely on the changes in the energy levels of electrons within the lasing medium include the argon-ion laser (pumped directly by electrical energy) which can work only as a continuous wave laser, and the neodymium-YAG laser. In the latter, the excited neodymium atoms are held in a crystal lattice of yttrium, aluminium and garnet. Light energy, either continuous or pulsed, 'pumps' the neodymium atoms.

The carbon dioxide laser works by taking advantage of the changes in the vibrational and rotational energy levels in the molecule that electromagnetic radiation of radio frequencies induces. The helium-neon laser is a gas-discharge laser, similar to the argon-ion laser, but its maximum output is around 5 milliwatts, so it is used as an aiming beam for the infrared lasers. A recent introduction is the excimer laser, which releases photons in the ultraviolet region of the electromagnetic spectrum.

Dye lasers are an exception to the single-wavelength devices. Here, the lasing medium is a complex organic dye. Stimulated emission can occur from complex molecules of this kind as a result of changes in their vibrational and rotational states, as well as from electrons moving between energy levels. These lasers have a band-width of about 100 nanometres. Individual wavelengths can be selected with a tuning device inside the cavity. Changing the dye will alter the wavelength more markedly.

The lasers used in medicine together with their operating characteristics and wavelengths, are listed in Table 7.1.

**This article is taken from 'Lasers take a shine to medicine', by Frank Cross, *New Scientist* p. 20.2.86.**

**TABLE 7.1** Laser – the coherent substitute for a scalpel

CONTINUOUS-WAVE LASERS

| LASER | WAVELENGTH/NM | POWER/W | FIBRE TRANSMISSION |
|---|---|---|---|
| $CO_2$ | 10 600 | 0.1–50 | No |
| Nd-YAG | 1064 | 0.5–100 | Yes |
| Argon | 488 514 | 1–10 | Yes |
| Dye | Tunable | 0.05–5 | Yes |

PULSED LASERS

| LASER | WAVELENGTH/NM | PULSE DURATION | ENERGY | FIBRE |
|---|---|---|---|---|
| Nd-YAG (QS) | 1064 | nanoseconds | 0.1–1J | No |
| Nd-YAG | 1064 | microseconds | 0.1–1J | Yes |
| Dye | tunable, visible | microseconds | 0.01–0.1J | Yes |
| Excimer | ultraviolet | nanoseconds | 0.01–0.1J | Yes |

Energy and power parameters are those used specifically in medical applications rather than the whole range obtainable from a particular laser.

## QUESTION

**7.5** Read the boxed article on the operation of medical lasers and answer the following questions:
   (a) Explain, with the help of a labelled diagram, the way that laser light is produced, and why the process is called Light Amplification by the Stimulated Emission of Radiation.
   (b) In the article, which 'properties suit the laser to medical purposes very well' and why?
   (c) How does Q-switching produce a pulsed output?
   (d) What is the difference of output between a pulsed and continuous-wave laser?
   (e) Which type(s) of laser are gas discharge? How is the energy supplied to these?

## Absorption of laser light

Tissue such as skin absorbs the far infrared light of the carbon dioxide laser in about 0.1 mm depth. Near infrared radiation from the neodymium-YAG laser is absorbed over distances of a few centimetres. The presence of the black pigment melanin in the skin, increases the rate of energy absorption. So too does any carbon which may be produced if the tissue is burnt. The haemoglobin in blood is a good absorber of the blue-green light of an argon laser.

The extent of the damage to tissue is reduced by a pulsed laser, compared to a continuous wave. There are difficulties in passing the higher power pulsed lasers down an optical fibre, since it can be shattered by thermal shock. The long wavelength of the carbon dioxide laser is absorbed by the fibre, so this cannot be used in a fibre optic endoscope.

The damage process in tissue is essentially due to the heating effect. The effect at high energy levels is to boil the water in cells, causing local damage to the cell and more widespread physical disruption as a result of the pressures produced by the steam generated. This is called photovaporisation. At lower energy levels proteins photocoagulate and change their chemical and physical nature, in a way similar to what happens in cooking egg-white.

## Uses of lasers

The laser has found a place in medicine for specific applications, its more general use being limited by the high cost of the equipment (scalpels are simple and cheap, if nasty!). Some examples of present uses are given here; in general they are operations that would be very difficult to carry out in any other way.

The carbon dioxide laser is used for cutting surface skin, and for delicate surgery on the brain and the fallopian tubes. The neodymium-YAG laser is used for internal removal of blockages. The laser beam is passed down an additional fibre in the endoscope, whilst the site is viewed with the image guide fibre. This laser can also be used to seal bleeding ulcers, by cauterising them and to remove fatty deposits (*atheroma*), from the blood vessels, as these can lead to heart failure.

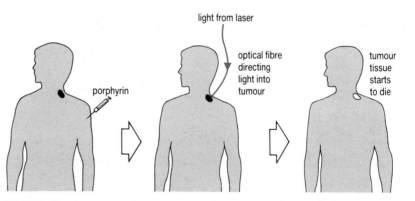

**FIG 7.11** Phototherapy: the photosensitiser, porphyrin, is injected into the body and given time to spread before light is passed down an optical fibre into the tumour. A few hours later, the cancer cells start to die.

The argon laser has been particularly successful in treating eye disease which results from diabetes. The laser is able to stop the bleeding which results, and prevent blindness. It is also able to repair small tears in the retina by photocoagulation. In a process which is referred to as 'spot-welding' a pulse of about 0.1 J of energy is delivered in less than 1 ms over an area of about $10^{-3}$ mm$^2$. Thus only a few cells are targeted and the operation is virtually painless. This laser is also used to remove unsightly birthmarks such as the port wine stain which results from having abnormal blood vessels at the surface of the skin. Present development work is concentrating on the possible use of lasers in the treatment of cancers. One approach is to inject into the body a chemical that is particularly sensitive to light – a photoactive substance, which the laser would then react with (in the tumour), making the chemical toxic to tissue. This technique called phototherapy is illustrated in Figure 7.11; it is beginning to emerge from the development stage into hospital use. More speculative is the possibility of producing a spot of laser light small enough to carry out surgery on a chromosome – destroying genes responsible for inherited abnormalities such as haemophilia and cystic fibrosis.

## SUMMARY ASSIGNMENTS

**7.6** Relate the uses of lasers that are described here to the mechanisms of their interaction with tissue, for example the use of argon lasers to destroy blood cells because this light is preferentially absorbed by red haemoglobin. (See Figure 7.1)

**7.7** This question compares the power of various lasers.
   **(a)** Calculate the intensity of the light used in 'spot-welding' the retina, in the example given above (in W mm$^{-2}$).
   **(b)** The laser you may have used in physics demonstrations is a helium-neon continuous-wave type with a maximum output of 1 mW and a minimum diameter of 1 mm.
      **(i)** What is the maximum intensity of light from this laser?
      **(ii)** If this intensity falls on the outer surface of the eye and the area of the spot is then reduced by $10^{-3}$ by the focusing of the cornea and lens, what is the maximum intensity on the surface of the retina?
   **(c)** Research has shown that the retina can be damaged by intensities as low as $10^{-2}$ W mm$^{-2}$. How does this compare with your answers to parts **(a)** and **(b)**?
   **(d)** Suppose that you accidentally allowed light from the He–Ne laser into your eye.
      **(i)** What should you do?
      **(ii)** What would you observe as a result of the exposure?

**7.8**

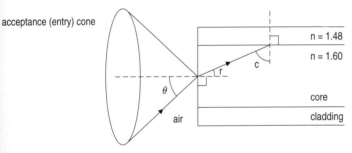

**FIG 7.12**

   **(a)** The optical fibre shown in Figure 7.12 has a glass core of refractive index 1.60 and glass cladding of refractive index 1.48. Calculate the critical angle, *c*, for a light ray travelling in the core of the fibre and being reflected at the core–cladding interface.
   **(b)** A light ray in air is shown entering the end of the optical fibre. Calculate the maximum angle $\theta$ of the entry cone in which rays can enter the end of the fibre and be totally internally reflected.

*(NEAB Mar 1997)*

**7.9** This question is about fibre optics and its application in endoscopy.
   **(a)** Figure 7.13 represents a section of a glass fibre with *cladding*.

**FIG 7.13**

(i) Draw Figure 7.13 and continue the path of the ray which enters the fibre at P and is contained within the fibre.

(ii) Label the angles of incidence and refraction at the entry point and relate them to Snell's law.

(iii) Explain the term *critical angle*.

(iv) Explain the condition under which the maximum angle of incidence is achieved for a ray which is contained within the fibre.

(v) State the meaning of *half-angle* for the fibre.

(vi) What is the function of the cladding?

(vii) State the necessary condition for the refractive index of the cladding.

Figure 7.14 shows a cross-section through the probe of a typical fibre-optic endoscope. The two light channels both contain fibre-optic bundles, only **one** of which is a coherent bundle.

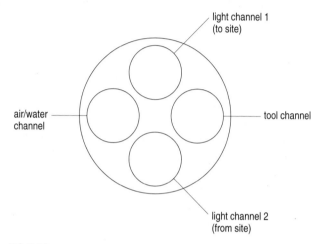

**FIG 7.14**

**(b) (i)** How are the fibres positioned in a coherent bundle?

**(ii)** State which of the light channels contains the non-coherent bundle in the endoscope. Explain why this bundle satisfactorily fulfils its task.

**(iii)** State the function of the coherent fibre-optic bundle in the endoscope and explain why a coherent bundle is required.

*(O & C June 1997)*

**7.10 (a)** State **two** physical properties of laser light which make it different from the light from a lamp bulb. Explain your answers.

**(b)** Give **two** examples of medical applications of a laser. In each case give a reason why the laser would be used.

*(AQA Mar 1999)*

**7.11 (a)** A medical laser operates at a beam power of 20 W with a beam diameter of 1.5 mm. The beam is focused by a convex lens to a spot of diameter 0.15 mm. Calculate the intensity of the radiation

**(i)** in the laser beam before focusing,

**(ii)** in the focused spot.

**(b)** With reference to your answers in **(a)** discuss the use of a medical laser as a scalpel.

*(OCR June 1999)*

# 8 ULTRASONICS

The increasing use of ultrasonics is immense. It is reckoned that the world wide market for scanners is in excess of £3 billion and that the investment in new developments is over £100 million.

The limit of human hearing is a frequency of about 20 000 vibrations per second (20 kHz), though animals make use of sound at considerably higher frequencies. Bats and dolphins, for example, emit pulses of sound in the 30–100 kHz range, and navigate by listening to the echoes. During the First World War Langevin discovered how humans could use these ultrahigh frequencies, though the technique was not sufficiently developed until the Second World War when it was used for locating underwater objects such as submarines. The emissions are called ultrasonic waves, or ultrasound and their field of use, ultrasonics, has developed greatly from this original aim. Ultrasonics offers a method of investigating objects internally without causing any damage. It is widely used in industry; in this chapter we will study its growing use in medicine.

## LEARNING OBJECTIVES

*After studying this chapter you should be able to:*

① use the following scientific terms correctly: piezoelectric, sonar, acoustic absorption, acoustic reflection, coupling medium;

② describe the production and detection of ultrasound by piezoelectric devices;

③ describe the process of ultrasound reflection by the body, and account for the formation of signals by tissue boundaries;

④ explain the principles of operation of the different types of ultrasound scanners: 'A','B','M', and 'real-time', and give examples of their use in medical diagnosis;

⑤ explain how the Doppler effect is used in the measurement of blood flow and foetal heartbeat;

⑥ describe the physiological effects of ultrasound on the body, and explain how these are used in therapy.

# GENERATION AND DETECTION OF ULTRASOUND

**8.1**

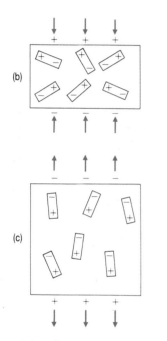

**FIG 8.1** The piezoelectric effect
**(a)** unstressed **(b)** compressed
**(c)** extended.

**T**he ultrasonic **transducer** is the device which can both generate and detect ultrasound vibrations. In principle a number of methods are possible, such as the use of magnetic effects. In medicine the **piezoelectric effect** is almost always the method used to convert electrical energy into ultrasound, and the reverse.

## The piezoelectric transducer

In a piezoelectric material, when an electric field is applied there is a change in the physical dimensions. This is caused by an interaction between the charges bound in the crystal lattice and the applied electric field. Conversely, if a mechanical stress is applied, a potential difference is produced across the ends of the crystal (Figure 8.1). The applied voltage can be 'shock excitation' or alternating, to cause a vibration of the crystal. The maximum transfer of energy will occur at the natural frequency of vibration of the crystal. This is **resonance**, which also ensures that the ultrasound energy is in a certain frequency range, which produces better images. To facilitate this the crystal has a thickness of one half of a wavelength of the ultrasound, producing resonance at the fundamental or lowest frequency.

The piezoelectric effect is present in a number of crystalline substances including quartz, lithium sulphate and barium titanate. The most commonly used substance is the artificial ceramic, lead zirconate titanate (PZT), which is a mechanically robust material which can be machined into a range of forms for different uses.

The transducer is made from a thin slice of the crystal, protected by a cover which usually incorporates a lens to converge the beam slightly (Figure 8.2). The beam consists of short pulses so the crystal is backed by epoxy resin which damps the vibrations quickly. The two faces of the crystal are covered with thin layers of silver as electrodes. One of these is connected to the earthed case which provides both mechanical and electrical protection. The other electrode is connected by a coaxial cable to (i) the power supply when acting as a transmitter, and (ii) an amplifier and cathode-ray tube when acting as a receiver.

**FIG 8.2** The piezoelectric probe.

## 8.2 ULTRASOUND IN THE BODY

Information about the internal structure of the body can be obtained by echo-location techniques using frequencies of several megahertz (MHz). The ultrasound is transmitted into the body where it is reflected at the boundaries between different types of tissue. It is also absorbed by tissue. So from the time lapse and relative intensity of the returning signal, the nature and position of the organs and tissues can be deduced. This technique is sometimes called sonar, **so**und **na**vigation and **r**anging.

### Properties of ultrasound

These sound waves behave in the same way as audible sound, which is described in Chapter 3. In particular the wave equation

$$c = f\lambda$$

applies. The speed of sound in soft tissues is about 1500 m s$^{-1}$ (Table 8.1), so for frequencies in the MHz range the wavelength is typically less than a millimetre. The basic echo-location method involves measuring the time interval between the emission of the pulse and its reception.

**TABLE 8.1** Ultrasound properties of materials

| MEDIUM | DENSITY ($\rho$) /kg m$^{-3}$ | ULTRASOUND VELOCITY ($c$) /m s$^{-1}$ | SPECIFIC ACOUSTIC IMPEDANCE Z (= $\rho c$) /kg m$^{-2}$ s$^{-1}$ |
|---|---|---|---|
| *Biological* | | | |
| Air | 1.3 | 330 | 429 |
| Water | 1000 | 1500 | $1.50 \times 10^6$ |
| Blood | 1060 | 1570 | $1.59 \times 10^6$ |
| Brain | 1025 | 1540 | $1.58 \times 10^6$ |
| Fat | 925 | 1450 | $1.38 \times 10^6$ |
| Eye | | | |
|   (aqueous humour) | 1000 | 1500 | $1.50 \times 10^6$ |
|   (vitreous humour) | 1000 | 1520 | $1.52 \times 10^6$ |
| Soft tissue (average) | 1060 | 1540 | $1.63 \times 10^6$ |
| Muscle (average) | 1075 | 1590 | $1.70 \times 10^6$ |
| Bone (varies) | 1400 | 2890 | $5.6 \times 10^6$ |
| | 1908 | 4080 | $7.78 \times 10^6$ |
| *Transducers* | | | |
| Barium titanate | 5600 | 5500 | $30.8 \times 10^6$ |
| Lead metaniobate | 5800 | 2759 | $16.0 \times 10^6$ |
| Lead zirconate titanate (typical) | 7650 | 3791 | $29.0 \times 10^6$ |
| Lithium sulphate | 2060 | 5437 | $11.2 \times 10^6$ |
| Quartz | 2650 | 5736 | $15.2 \times 10^6$ |

QUESTION

**8.1** **(a)** With reference to Table 8.1, calculate for an ultrasound wave of 1.5 MHz:
   **(i)** the wavelength in bone, fat and the piezoelectric crystal PZT,
   **(ii)** the thickness of a PZT crystal in a transducer, producing its fundamental resonance.
**(b)** The time delay for an echo from ultrasound in soft tissue was 0.133 milliseconds. At what depth was it reflected?

## Reflection

This occurs at a boundary because of a difference in the characteristic **acoustic impedance** Z, of each substance. This is simply the product of the density $\rho$ and the velocity of sound $c$ through it. As Table 8.1 shows, the impedance is similar for most soft tissues, so the amount of energy reflected at their boundaries is small, typically less than 1 per cent. It can be shown that the ratio of the reflected intensity $I_r$ to the incident intensity $I_i$ is:

$$\frac{I_r}{I_i} = \frac{(Z_2 - Z_1)^2}{(Z_2 + Z_1)^2}$$

This is called the **intensity reflection coefficient** $\alpha$. Clearly there will be no reflection if the impedance of the two media is the same, $Z_1 = Z_2$.

QUESTION

**8.2** **(a)** What is the value of $\alpha$ going from bone to soft tissue?
**(b)** What will happen at the interface between the lung (full of air) and the surrounding tissue?

The answer to **8.2(b)** shows that it is important to ensure there is no air between the transducer and the skin. This is done by interposing a **coupling medium** such as a gel or oil. In a similar way the presence of gas anywhere in the field of view of the ultrasound will cause a problem for imaging.

A reflection of 1 per cent of the incident energy at a soft tissue boundary is sufficient for a sensitive detector, and this has the advantage that the remainder of the beam can penetrate more deeply into the body, so that reflections from a series of tissue boundaries can be detected (Figure 8.3).

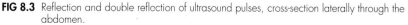

**FIG 8.3** Reflection and double reflection of ultrasound pulses, cross-section laterally through the abdomen.

## Resolution and attenuation

Resolution is a measure of the smallest detail that can be detected, and so is an important feature of all imaging systems (see Section 2.2 and 2.3 for example). It depends on the wavelength of the ultrasound, the smaller this is the better the resolution, rather like using a finer point to a brush to paint smaller features. This means that higher frequency ultrasound gives more detailed pictures.

The ultrasound beam spreads out as it passes through tissue, and is scattered and absorbed by the molecules it meets. Unfortunately this attenuation also increases as the frequency increases. In practice there has therefore to be a compromise between resolution and penetration, in the choice of frequency for any particular application.

### INVESTIGATION

#### *The properties of ultrasound*

Ultrasound transmitters and receivers are produced for laboratory use, for investigating their wave properties, for example, Ultrasonic kit from Unilab. If this equipment is available to you, follow the suggested methods of investigating the reflection, refraction and interference of ultrasound waves. Try to devise an arrangement to simulate the passage of ultrasound through different media (e.g. water and bone) to investigate acoustic coupling.

Alternative sources of ultrasound are the motion sensor from Educational Electronics, or you could try your local hospital for redundant equipment.

## 8.3 SCANNING AND IMAGING

The pulse echo received from the body is converted by the transducer into an electrical signal which is converted to digital information to be stored by computer, displayed on the screen of a cathode-ray tube and sometimes recorded on videotape. There are several different modes of display available for different purposes.

### A-scan

This is the range-finding display, in which the horizontal axis represents time, corresponding to the distance into the reflecting tissue, and the vertical axis represents the amplitude of the echo. Figure 8.4 shows how this works. The first pulse received is from the abdominal wall. The time for the pulse to echo is a function of the thickness of the wall, and so the A-scan could be used to measure it – assuming that the clinician knew the anatomy of the body section being scanned! Likewise the two boundaries of the organ give pulses of decreasing size because of attenuation, whereas the bony vertebra at the back gives a higher intensity because it reflects a greater proportion of the sound.

A-scanning is useful in those situations where the anatomy of the section is well known, and a precise depth measurement is required. One such example is in measuring the long axis of the eye before a cataract operation. It is also used in dermatology for determining skin thickness and depth of a burn. For most purposes however, the **a**mplitude-modulated **A**-scan information can be more usefully presented as a **b**rightness-modulated display called a **B**-scan (Figure 8.4). This is the basis of several types of display.

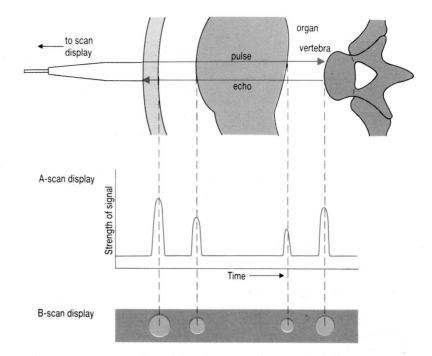

**FIG 8.4** Pulse display by amplitude (A-scan) and brightness (B-scan).

(a)                                                                    (b)

**FIG 8.5 (a)** M-scan showing abnormal mitral valve and left ventricle movements **(b)** M-scan of same patient after corrective treatment.

## M-scan (motion or time-position scan)

This is a useful technique for monitoring the regular movement of an organ such as the heart. It is a combination of the A-scan and the B-scan. The dots are produced in positions along the x-axis, according to the depth of the reflection. The Y-plates are connected to a time-base so that the dots move down the screen at a low speed. A change in the position of the interfaces, such as the heart wall, will therefore

(a)

probe

placenta

body
cross-
section

limbs

foetal skull

spine

(b)

(c)

(d)

(a) trace      (b) trace   (c) trace

**FIG 8.6** Building up the B-scan image.

result in horizontal displacement of the dots and a non-straight line. A regular oscillation will produce the wavy trace shown in Figure 8.5. The viewing of the heart is limited by surrounding bone and air, so care has to be taken to position the probe to avoid the ribs and lungs. In addition the heart valves are not plane surfaces so do not produce strong reflections. The technique of ultrasound cardiography (UCG) is therefore one which requires considerable skill.

## Two-dimensional B-scan

This is particularly important in medical diagnosis as it results in the production of a pictorial display of the organs, at a particular cross-section. In order to produce these the transducer probe is moved so that the body is viewed from a range of angles. These different views give rise to a set of brightness points. These are automatically correlated with information about the position and orientation of the probe to build up the two-dimensional picture (Figs 8.6, 8.7). The B-scan technique has proved most valuable in obstetrics where it is now a routine procedure for monitoring the health of the pregnant woman and the growth of the foetus. This latter is usually done by accurately measuring the widest diameter of the skull, the bi-parietal diameter (see Figure 8.6). Various complications of pregnancy such as faults in the amniotic fluid and the placenta, can be diagnosed early; later scans can show any developmental problems in the foetus and check the position before birth.

Two-dimensional images are also used for detecting cysts, abscesses and tumours in organs such as the liver, kidney and ovary. They are invaluable in the examination of aneurysms, that is weak, ballooned-out parts of arteries. Follow-up to these investigations may be complementary X-ray or radioisotope imaging, radiotherapy or surgery. In addition there is a growing use of this type of imaging to assist the surgeon with minor operations through the skin with very small incision. Cysts can be drained and minor blood vessels repaired quickly and without major trauma to the patient.

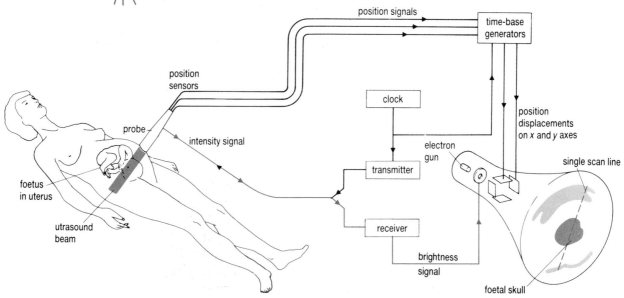

**FIG 8.7** The mechanism of B-scanning. Position sensors in the probe send signals to the X- and Y-plates of the CRT. The received ultrasound reflection is fed to the electron gun to produce a change in the brightness.

**QUESTION**

**8.3** With reference to the description and illustrations of this section, write an account of how a two-dimensional image of varying brightness is produced. Include in your notes a simple block diagram to represent the required instrumentation, and discuss the merits of the different scanning movements.

## Real-time B-scans

Most hospital ultrasound equipment now incorporates real-time scanning, which is particularly useful when the body structures are moving (which is quite often, and almost always in the case of obstetric examinations!). Real-time systems take a series of pictures in very rapid sequence, rather like cine filming. Although the detail of any one picture is reduced by the speed at which it was taken, seeing the sequence reveals more than the best still picture. Movement brings features to life so that diagnosis with a real-time scanner is easier, requiring less manipulative skill and detailed anatomical knowledge. They rely on an **array** of transducers and use computer control to operate this in rapid sequence, and to process the received pulses to give frame rates of up to 150 per second (Figure 8.8). These systems are able to combine a recently stored image with the current real-time image to improve resolution.

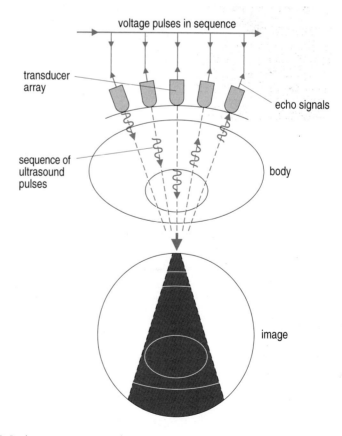

**FIG 8.8** Real-time scan using transducer array.

**FIG 8.9** Doppler effect **(a)** stationary source and observer **(b)** moving source **(c)** moving observer.

**M**easurement of a continuous movement such as the flow of blood, can be made using the effect of the moving reflector on the frequency of the ultrasound. This change is called the Doppler effect. You will have noticed this even though you may not have realised what it was. When a siren on a vehicle moves past, you hear a sudden fall in the pitch. The frequency of the siren note was higher when it approached you than when it receded from you. The effect is used in police 'radar traps' to measure whether vehicles are travelling faster than the speed limit. These use electromagnetic radiation, for the effect is found in all types of wave. We will see how this occurs.

## The Doppler effect

**Moving source** Suppose the source of sound has a frequency $f_s$ in a medium in which the wave speed is $c$. In one second $f_s$ waves will be emitted and will occupy a distance $c$ (Figure 8.9(a)). If however, the source is moving at a speed of $v_s$ towards the observer, then the $f_s$ waves will be compressed into a distance $c-v$ (Figure 8.9(b)). Let us see what effect this has on the frequency. The observed wavelength $\lambda_o$ is the distance divided by the number of waves:

$$\lambda_o = \frac{c - v_s}{f_s}$$

so the observed frequency

$$f_o = \frac{c}{\lambda_o} = \frac{cf_s}{c - v_s}$$

which is higher than the source frequency.

This change from the original frequency is called the Doppler shift, in this case:

$$\Delta f_1 = f_o - f_s = \frac{cf_s}{c - v_s} - f_s$$

simplifying

$$\Delta f_1 = \frac{f_s v_s}{c - v_s}$$

In the case of ultrasound and blood flow, $c \gg v_s$ so this further simplifies to

$$\Delta f_1 = \frac{f_s v_s}{c}$$

**Moving observer** Suppose the observer moves towards the stationary source with speed $v_o$ (Figure 8.9(c)), then the wavelength remains $\lambda_s$ but the relative speed has increased to $v_o + c$, so that the observed frequency becomes

$$f_o = \frac{c + v_o}{\lambda_s} = \frac{(c + v_o)}{c} f_s$$

In this situation the Doppler shift is

$$\Delta f_2 = f_o - f_s = \frac{(c + v_o)}{c} f_s - f_s$$

$$\Delta f_2 = \frac{f_s v_o}{c} \tag{1}$$

**Ultrasound reflector** Let us now apply this to the ultrasound wave frequency $f$, reflecting off a surface which is moving towards the detector with speed $v$ (Figure 8.10). The total Doppler shift is made up of two relative motions:

(i) the surface as observer, is moving towards the transmitter source so receives a Doppler shifted frequency $f'$ from equation 1 above,

$$\Delta f_2 = f' - f = \frac{fv}{c}$$

(ii) the surface as source, is moving towards the receiver as observer, which results in the Doppler shifted frequency $f''$, from equation 1 above,

$$\Delta f_2 = f'' - f' = \frac{fv}{c}$$

Therefore the sound has a total Doppler shift which is double the one-way shift:

Total Doppler shift $\Delta f = \dfrac{2fv}{c}$

For example, a 2 MHz ultrasound beam travelling at 1.5 km s$^{-1}$ will have a Doppler shift of about 270 Hz from blood flowing at a speed of 0.1 m s$^{-1}$.

**FIG 8.10** Doppler shift from a moving reflector.

## Continuous wave Doppler system

The frequency of the ultrasound is chosen to give the optimum balance between resolution and penetration. For measurements near the surface the range used is 5–10 MHz, for the centre of the body it falls to 2 MHz. The beam for ultrasound is a continuous wave, unlike the pulse used in previously described methods. This means that the system cannot be used for depth measurement, and that it requires a separate *transmitter* and *receiver*. The signals from both are *electronically mixed* and the output is *filtered* so that only the Doppler-shift frequency is passed on to be *amplified*. These frequencies are usually in the audio range, so the operator wears *earphones* to 'hear' the moving structure. The tone will vary with the shift frequency, and so with the speed of the movement. If the structure is the beating heart then the tone will vary with the heartbeats. The operator may work completely on this 'real-time' effect, recognising characteristic pitches or patterns, but there is a *recording* facility available.

---

**QUESTION**

**8.4**  Draw a block diagram to show the process in a Doppler system (these are in *italic* in the previous paragraph).

---

## Uses

**Foetal heart monitoring**  This is the most common use of Doppler-shift ultrasound. The instrument is actually being used as a highly directional and sensitive stethoscope, though the sound that is heard is like galloping horses, which might alarm the expectant mother! It is normally possible to hear the foetal heart after the twelfth week; checks after that can determine not only whether the foetus is alive, but whether the foetus is in any difficulties, particularly during labour. This monitoring also includes the motion of the placenta and umbilical cord.

**Blood flow measurement** In most cases the blood is travelling at an angle $\theta$ to the direction of the beam, so the frequency shift is given by:

$$\Delta f = \frac{2fv}{c} \cos \theta$$

as shown in Figure 8.11. In this case it is necessary to measure $\theta$, (which will depend on the inclination of the probe relative to the blood vessel), as well as $\Delta f$. A blockage due to a clot (*thrombosis*), or a constriction due to plaque on the vessel walls (*atheroma*), is immediately apparent as a sudden change in the shift frequency. More detailed investigations can analyse the range of shifts produced by the reflection. This arises from the variation of the blood flow speed across the diameter of the vessel, usually parabolic in distribution, as shown in Figure 8.11 inset. Small obstructions will alter this pattern, so affecting the reflection. One problem is that arteries and veins often lie close together, so the two signals can interfere with each other.

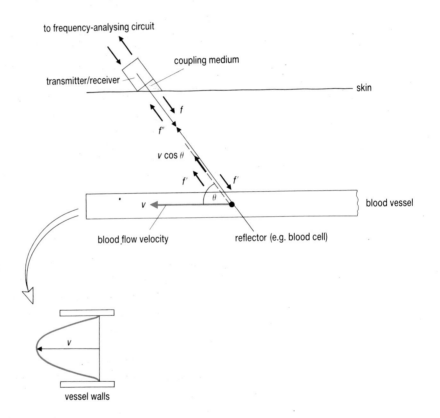

**FIG 8.11** Doppler blood flow measurement.

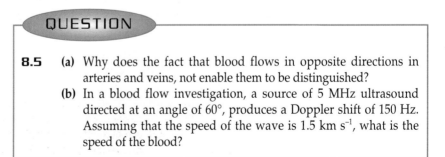

### QUESTION

**8.5** **(a)** Why does the fact that blood flows in opposite directions in arteries and veins, not enable them to be distinguished?

**(b)** In a blood flow investigation, a source of 5 MHz ultrasound directed at an angle of 60°, produces a Doppler shift of 150 Hz. Assuming that the speed of the wave is 1.5 km s$^{-1}$, what is the speed of the blood?

# PHYSIOLOGICAL EFFECTS OF ULTRASOUND

Ultrasound is a very safe way of investigating the interior of the body, hence the rapid growth of its use in recent years to replace X-rays. As a high frequency vibration of the body it can produce physical and chemical changes. These depend on the intensity of the beam; in diagnostic use this is very low (usually below 0.1 W cm$^{-2}$), much higher intensities (up to $10^3$ W cm$^{-2}$) are used for therapy.

## Damage mechanisms

Ultrasound waves interact with the thermal motion of the molecules of body tissue. At power levels of 1 W cm$^{-1}$ and above, this results in measurable **temperature** increase. An increase of more than 1.5 K is considered hazardous because it can cause chemical changes. The destruction of tissue can occur at levels of around 1000 W cm$^{-1}$.

The **pressure** changes between the compressions and rarefactions are the second cause of damage. The rapid physical movement can rupture tissues, just as the ear drum can be torn by an intense external source of sound. Pressure changes can also cause dissolved gases, e.g. $O_2$ and $CO_2$ in the blood to come out of solution and form bubbles. These bubbles collapse violently as the pressure changes. This release of energy can cause both **physical damage** and **chemical change**. For example water can be changed into $H_2$ and $H_2O_2$ and DNA molecules can be broken. Free radicals can result in oxidation reactions which disrupt the metabolism of that part of the body. These chemical effects are similar to those resulting from ionising radiations described in Theme 3; the cause is very different however.

## Ultrasound therapy

**Diathermy** Ultrasound is used to cause local heating of tissue deep inside the body in a similar way to that produced by microwave and electrical methods. The aim is to produce relief from pain for example in the joints of an arthritis sufferer. Typical treatments use intensity levels of several W cm$^{-2}$ for periods of a few minutes several times a week. The transducer is moved over the surface of the joint during treatment to avoid 'hot spots' developing. Ultrasound is particularly useful as a 'deep heat' source as it does not deposit much of its energy in the soft tissues near the surface of the body. It is not used in delicate organs such as the eye or the gonads.

**Tissue destruction** At intensity levels of 1000 W cm$^{-2}$ it is possible to selectively destroy tissue by focusing the beam on the required place. Investigations have shown that ultrasound can destroy cancer cells; recent advances have been reported in research into the treatment of prostate cancer. Ultrasound has also been successful in breaking up gallstones, avoiding the need for surgery.

---

### QUESTION

**8.6**  (a) Over what distance in soft tissue, does the maximum pressure change occur, for a 1 MHz wave?

   (b) Make a table of intensity levels of the various uses of ultrasound, and of the levels at which types of physiological effects occur.

## 8.6 DEVELOPMENTS IN ULTRASOUND

In recent years ultrasound has expanded rapidly in the range and extent of its applications in medicine. The main reasons for this are that it is inexpensive, non-invasive and, at the intensity levels required for diagnosis, no harmful effects have been detected. It is particularly good at imaging soft tissue, compared with radiological methods. These advantages have resulted in it being the only imaging technique used in obstetrics. It is also very important in gynaecology, cardiology and studying the vascular system. Ultrasound is now routinely used for almost all organs in the body (often in conjunction with other imaging techniques described in the following chapters).

Developments in ultrasound equipment have made possible much of this increased use. Improved resolution of the image is enabling foetal hearts to be examined for congenital defects earlier in the pregnancy. The electronic processing of the image has made it possible to compare changes in an image of a tissue over time to study, for example, the degeneration of a muscle in muscular dystrophy, so that prognosis and treatment can be determined. Electronic control of the transducer arrays allow more rapid scanning and the availability of relatively inexpensive, portable, real-time scanners is now very general (Figure 8.12).

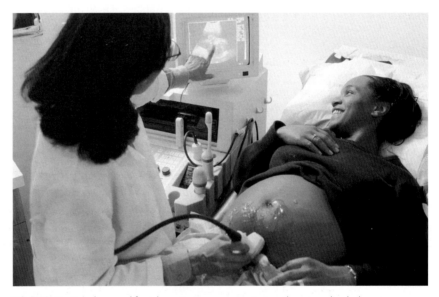

**FIG 8.12** Routine ultrasound foetal monitoring: an expectant mother views her baby on screen.

Measurements of Doppler flows can be displayed, using electronic analysis, on a CRO as a sonogram or spectral Doppler display. This shows in detail the variation of blood flow with time (Figure 8.13). Duplex scanning can present the Doppler waveform of the blood flow superimposed on a two-dimensional image of the blood vessel under scrutiny. Doppler is an inexpensive, though time-consuming screening technique for the ever-growing need to investigate the circulation system, with a view to surgery. 'Keyhole' surgery can be carried out at sites close to the surface of the body by using real-time ultrasound imaging whilst operating. An example is percutaneous transluminal angioplasty, which just means 'replacing part of an artery whilst looking through the surface at it'. A part of the artery which is damaged is located by imaging and a balloon catheter (Figure 8.14) is inserted through a small incision made under local anaesthetic. The balloon is then inflated until the artery is restored to its original shape.

**FIG 8.13** Patient undergoing the Doppler procedure and display of blood flow.

**FIG 8.14** Inserting a catheter for angioplasty.

## SUMMARY ASSIGNMENTS

**8.7** **(a)** The formula for the reflection coefficient for ultrasound is $[(Z_2 - Z_1)(Z_2 + Z_1)]^2$. What quantity does Z represent?

In a pregnant woman, the bladder is between the outside of the body and the baby. A pregnant woman needs to have a bladder full of urine if she wishes to have a successful ultrasound scan of her baby. The principal contents of an 'empty' bladder are gaseous.

With reference to the formula for reflection coefficient, explain why an 'empty' bladder would make an ultrasound unsuccessful.

Give one reason why ultrasound might be preferred to a method involving a radioisotope for investigating the size of a body organ.

**(b)** An ultrasonic transducer or probe typically produces pulses of 1 μs duration. The pulses are sent into the medium and the returning echoes are detected by the transducer, amplified and recorded. In general, the higher the frequency the better the resolution gained in diagnostic techniques. You would therefore expect that the highest possible frequencies would be used. Unfortunately, the higher the frequency, the greater the absorption of energy from the beam. Because of this, investigations near the surface all use a high frequency probe operating at 10 MHz, and fine detail can be seen. Investigations of deeper tissues use a lower frequency, e.g. 3.6 MHz.

[Adapted from G Hart and F Armes: *Medical Physics for Advanced Level*, Simon and Schuster Education 1992.]

    **(i)** Express 1 μs and 3.5 MHz in standard form, i.e. using powers of ten.
        Calculate how many times in 1 μs a 3.5 MHz ultrasonic probe moves in and out.
        Draw a sketch graph of the wave pulse produced in 1μs. Add a scale to the time axis.
    **(ii)** How is frequency related to wavelength for ultrasonic waves in one medium?
        Explain the phrase *'the higher the frequency the better the resolution'*.
    **(iii)** Give an example of an investigation which would be near the surface of the body.

*(Edexcel specimen 2001/2)*

**8.8** **(a)** A piezoelectric ultrasound transducer is made from a thin slice of an artificial ceramic such as lead zirconate titanate (PZT).

    **(i)** Describe what happens when an alternating voltage is applied to a PZT transducer so that ultrasound is produced.
    **(ii)** Under what conditions will maximum energy conversion into ultrasound occur?

**(b)**

**FIG 8.15**

The diagram Figure 8.15 shows the oscilloscope display of pulse amplitude against time for an ultrasound A-scan through a person's abdomen. Assume that the weaker echoes come from internal organs.
**(i)** Describe the procedures which are used to obtain this type of scan.
**(ii)** Explain how the spacing of the pulses is interpreted.
**(iii)** Give **two** reasons why the amplitude of the reflected pulses varies.
**(iv)** If the speed of ultrasound through water and soft tissue is about 1500 m s$^{-1}$, estimate the distance between the front of the patient's abdomen and the spinal column.

*(AQA Specimen 2001/2)*

**8.9** This question is concerned with the use of ultrasound in medicine.
**(a)** Explain the generation and detection of ultrasound.
**(b)** Describe the reflection of ultrasound at different interfaces, and the transmission of ultrasound through different media.
**(c)** Figure 8.16 illustrates an ultrasound scan of a bone, soft tissue and skin. Use the data below to describe in detail the structure being scanned.

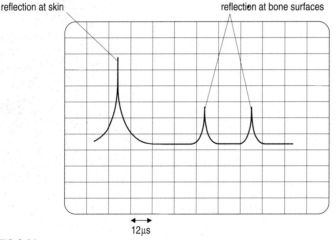

**FIG 8.16**

speed of ultrasound in soft tissue = 1500 m s$^{-1}$
speed of ultrasound in bone     = 4000 m s$^{-1}$

*(OCR Mar 1999)*

**8.10** **(a)** **(i)** Describe the technique used to monitor the signal produced by a Doppler-shift ultrasound machine.
   **(ii)** Suggest a typical medical condition or problem which could be investigated by this technique.
**(b)** Ultrasound, of frequency 4.0 MHz and speed in tissue of 1.5 km s$^{-1}$, is used to measure the speed of blood flow in an artery. The angle $\theta$ between the artery and the direction of the ultrasound source is 45°. If the Doppler shift in frequency is 800 Hz, calculate the speed of the blood in the artery.

The Doppler shift is given by $\Delta f = \dfrac{2fv \cos \theta}{c}$ .

*(NEAB Mar 1997)*

# IONISING RADIATIONS AND MAGNETIC RESONANCE IMAGING

**R**oentgen's discovery of X-rays in 1895 immediately captured popular imagination. His first paper included a radiograph of his wife's hand. Within a few months, the diagnostic use of X-rays was widespread. Many of the other claims for the new rays, including their potential for abolishing vivisection, smoking and alcoholism, were not sustained. Nor was their potential for damage to the body recognised for some time, despite the impression given by a contemporary cartoon.

This theme considers the effects of X- and other ionising radiations on the body. Examples of diagnostic and therapeutic use are presented, and the controversial issue of suitable safety precautions is reviewed.

Almost one hundred years later the development of MRI magnetic resonance imaging has had a similar impact on medical practice and popular perception. It is an expensive technique but, so far, appears to be completely safe. ■

Cartoon from *Life* 17 February 1896.

## PREREQUISITES

Before starting this theme you should have some familiarity with the following:

- The nature and properties of X-, alpha, beta and gamma radiations.
- The notation used in writing nuclear reactions.
- The decay and half-life of radioisotopes.
- Magnetic fields.

# 9 X-RAYS

You will probably have had at least one X-ray examination, by the time you read this book, as X-rays are used routinely for diagnosis in dentistry and many branches of medicine (Figure 9.1). In this chapter you will study the nature and uses of X-rays in some detail. We will look first at the nature of X-rays, and how this depends on their production, then how X-rays interact with tissue. This is followed by a section on the equipment which generates, controls and detects the X-ray beam. Finally we will review the range of uses of X-rays in diagnosis and therapy.

**FIG 9.1** X-radiograph of the teeth of a child, showing some second teeth below the first or milk teeth.

## LEARNING OBJECTIVES

*When you have studied this chapter you should be able to:*

① use the following terms correctly: photon energy, hardness, total linear attenuation coefficient ($\mu$), positron, fluoroscopy, tomography;

② recall the nature of X-radiation and the intensity–wavelength spectra produced by a tungsten target and their dependence on tube voltage and current and on filtration;

③ describe the mechanism of interaction of X-rays with the human body, and list some of the resulting types of damage;

④ use the relationships:
   (a)   Intensity in vacuum    $I = I_0/r^2$ (inverse square law)
   (b)   Intensity in a medium  $I = I_0 e^{\mu x}$
   (c)   Half value thickness

$$x_{1/2} = \frac{\log_e 2}{\mu}$$

⑤ describe the principle of production of X-rays by a rotating anode tube and the formation of a radiograph of body tissue;

⑥ Give examples of the use of X-rays in diagnosis and therapy.

# 9.1 THE NATURE OF X-RADIATION

## *Origin*

X-radiation is electromagnetic radiation of high photon energy. It is produced when charged particles, usually electrons, moving at high speed are slowed down rapidly by striking a target.

Only about one per cent of the energy of the electrons is converted to X-rays, the remainder being dissipated as substantial heating of the target. X-radiation is always emitted with a range of wavelengths, corresponding to the range of photon energies produced. There are two mechanisms by which this occurs: these are illustrated in Figure 9.2. In (a) an electron is slowed which results in Bremsstrahlung radiation, which is German for braking radiation. In (b) an inner electron is removed causing ionisation, and subsequently an X-ray is emitted.

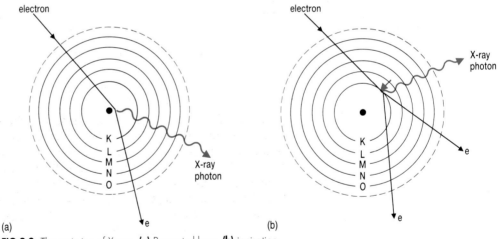

**FIG 9.2** The emission of X-rays **(a)** Bremsstrahlung **(b)** ionisation.

## *Spectra of X-radiation*

**Bremsstrahlung** radiation results in a continuous spectrum. There is a short wavelength limit to the spectrum, which is when all the energy of the incident electron is converted to a single X-ray photon. This energy

$$E_{max} = eV_0$$

where $e$ = the charge on an electron

$V_0$ = the peak value of the applied voltage

From the quantum theory

$$E_{max} = hc/\lambda_{min}$$

Where $h$ = Planck's constant, $c$ = the velocity of light and $\lambda_{min}$ = the minimum wavelength

Therefore the short wavelength limit is given by;

$$\lambda_{min} = hc/eV_0 = 1.24 \times 10^{-6} \text{ m/peak voltage (in volts).} \tag{1}$$

This important relationship is sometimes known as the **Duane-Hunt Law**. The long wavelength limit of the spectrum is not so clearly defined. Being of low energy, these X-rays are easily absorbed by the X-ray generator.

**Ionisation** results in the **characteristic** or **line** spectrum. Each line is the result of the change in energy of a particular electron in an atom of the target. The lines are named according to the shell in which the electron terminates. Those with the shortest wavelength which result from a transition to the innermost K-shell, are K-lines. The applied voltage must be above a critical value for this to occur. For tungsten, the usual target material, this value is about 70 kV, for K-lines. A typical spectrum is shown in Figure 9.3.

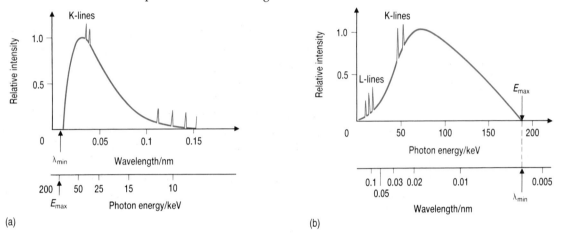

**FIG 9.3** Typical X-ray spectrum **(a)** intensity vs wavelength **(b)** intensity vs photon energy.

## The intensity and quality of X-rays

The effects of X-radiation depend both on the intensity (energy per unit area), and on the spectral spread. It is the latter which determines how penetrating the radiation is, and is called the **quality**. To describe this fully requires a graph such as Figure 9.4. In practice however, it is usually sufficient to specify the quality in simpler terms, for example, the thickness of material to reduce the intensity to a half. This is called the **half-value thickness** and is explained in the next section.

We will now consider the factors affecting the intensity and quality of the X-ray beam, resulting from changes in the Bremsstrahlung and characteristic radiations.

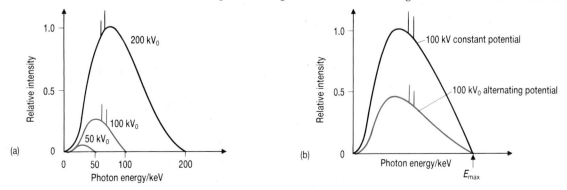

**FIG 9.4** Effect of tube voltage on spectrum **(a)** change of peak voltage **(b)** alternating power supply.

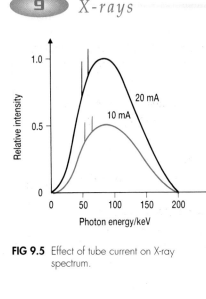

**FIG 9.5** Effect of tube current on X-ray spectrum.

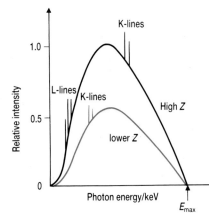

**FIG 9.6** Effect of target material on X-ray spectrum.

**Tube voltage** As the tube voltage is increased the following changes occur, as shown in Figure 9.4(a):

■ $E_{max}$ increases and $\lambda_{min}$ decreases;

■ the peak intensity of the continuous spectrum moves to higher energies;

■ the total intensity, which is given by the area under the curve, increases rapidly; it is proportional to $V_0^2$;

■ more characteristic lines may appear in the spectrum.

In addition the beam will be affected by variations in the supply of the tube voltage over time. Some X-ray generators have an alternating power supply which results in a greater proportion of low energy radiation (Figure 9.4(b)).

**Tube Current** As the tube current is increased, more electrons are emitted from the cathode, which results in the following changes (Figure 9.5):

■ $E_{max}$ remains unchanged, as $V_0$ is constant;

■ the shape of the spectrum remains unchanged;

■ the total intensity increases; it is proportional to the tube current.

**Target material** As the proton number Z of the target material increases, the following effects are observed (Figure 9.6):

■ $E_{max}$ remains constant;

■ the total intensity increases because there is a greater probability of collision between the bombarding electrons and the larger, more positively charged target nuclei;

■ the characteristic line spectra are shifted to higher photon energies.

The target material needs to have a high proton number Z, but also a high melting point because of the heat generated. Tungsten is the ideal material, as it has Z = 74 and a melting point of 3650 K.

**Filters** A thin sheet of material placed in the beam, will selectively absorb more low energy photons than high energy photons, resulting in a change of the kind shown in Figure 9.7. The filtered beam is more penetrating, and said to be **harder**.

## QUESTIONS

**9.1** List the changes in the quality and intensity of the beam by filtration, shown in Fig 9.7.

**FIG 9.7** Effect of filtration on X-ray spectrum.

**9.2** Construct a table to summarise the effects of $V_0$, $I_0$, and Z of the target, on $E_{max}$, $\lambda_{min}$, the shape of the spectrum, the characteristic lines and the total intensity.

**9.3** Sketch Figs 9.5 and 9.6 with $\lambda$ as *x*-axis instead of photon energy (as in Fig 9.3(a)). Note that you can use a conversion factor from equation 1, which simplifies to:

wavelength in nanometres = 1.24/energy in keV.

**9.4** An X-ray tube is operated with a peak voltage of 100 kV; the beam current is 40 mA.
(a) At what rate does the machine transform energy?
(b) How many electrons reach the target each second?
(c) What is the maximum energy of the X-rays produced?
(d) What is the minimum wavelength of these X-rays?

$$e = 1.6 \times 10^{-19}\,\text{C},\ h = 6.6 \times 10^{-34}\,\text{J s},\ c = 3 \times 10^8\,\text{m s}^{-1}$$

# THE INTERACTION OF X-RAYS WITH MATTER

**9.2**

## *The sequence of events*

This is shown in summary in Figure 9.8.

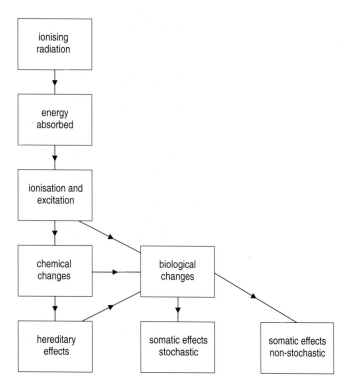

**FIG 9.8** Absorption of radiation by body tissue.

When a beam of X-rays passes through matter, such as body tissue, energy is given to the tissue and the intensity of the beam is reduced, or **attenuated**. In the interaction, some of the energy is **scattered**, and some is **absorbed**. Absorption can cause **excitation**, in which an electron is raised to a higher energy level within the atom or molecule, or **ionisation**, in which the electron is removed. This ionisation may damage or destroy biologically important molecules directly. Alternatively, this may occur indirectly through chemical changes in the surrounding medium, which is mainly water.

For example, the ionisation of water molecules produces O and OH free radicals, which have unpaired electrons and are highly reactive. ($H^+$ and $OH^-$ ions which are also produced, are present in ordinary water and are not particularly reactive). Two OH radicals may react to produce $H_2O_2$, the powerful oxidising agent hydrogen peroxide, which can then attack the complex molecules which form the chromosomes in the nucleus of each cell.

A range of biological effects can follow this. These are summarised in Table 9.1. They can be divided into two classes, **somatic** and **hereditary**. Somatic (meaning of the body) effects arise from damage to the ordinary cells of the body and affect only the person irradiated. Hereditary effects arise from damage to the reproductive organs and so damage can be passed on to the person's children.

**TABLE 9.1** Biological effects of radiation

| LEVEL OF BIOLOGICAL ORGANISATION | IMPORTANT RADIATION EFFECTS |
|---|---|
| Molecular | Damage to macromolecules such as enzymes, RNA and DNA, and interference with metabolic pathways. |
| Subcellular | Damage to cell membranes, nucleus, chromosomes, mitochondria and lysosomes. |
| Cellular | Inhibition of cell division; cell death; transformation to a malignant state. |
| Tissue; organ | Disruption of such systems as the central nervous system, the bone marrow and intestinal tract may lead to the death of animals; induction of cancer. |
| Whole animal | Death; 'radiation lifeshortening'. |
| Populations of animals | Changes in genetic characteristics due to gene and chromosomal mutations in individual members of the species. |

The presence and severity of these effects depends of course on the amount of energy absorbed by the irradiated tissue. This will be considered in Chapter 11. A distinction is made between **stochastic** and **non-stochastic**. Stochastic effects are those for which there is always some possibility of the effect occurring; the probability increases with the dose of radiation received. Non-stochastic effects occur only above a certain threshold, then the severity of the effect increases with the dose of radiation received. Hereditary effects are stochastic and can be seen in the statistics for large populations (because the probability is small). Somatic effects can be either stochastic, e.g. leukaemia, or non-stochastic, e.g. burns of the skin.

## Attenuation mechanisms

Four processes can result in loss of energy from an X-ray beam to soft tissue. They are shown in order of increasing photon energy in Figure 9.9.

**Scatter** The energy of the X-ray photon, $E$ is smaller than the energy required to remove an electron from its atom (this is the binding energy $E_b$).

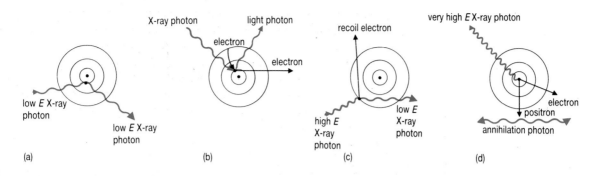

**FIG 9.9** The four processes of attenuation **(a)** simple scatter **(b)** photoelectric effect **(c)** Compton scatter **(d)** pair production.

**Photoelectric effect** The photon, with energy slightly greater than $E_b$, ejects an electron from its atom. Another electron in the atom then 'drops into' the vacancy with the emission of characteristic photons.

**Compton scatter** The photon has much more energy than the binding energy, so that a photon of reduced energy is scattered from the interaction with the ejected electron, which is known as a recoil electron.

**Pair production** At very high photon energies, the photon can interact with the nucleus of an atom. The photon disappears and an electron and a positron emerge. These two particles lose their energy by ionisation, until a positron is annihilated by an electron, with the generation of two identical photons.

> A **positron** is a fundamental particle, identical to the electron, but with an opposite charge.

## Attenuation and distance

If the energy of the X-rays radiates from the source in all directions, the intensity will fall in proportion to the square of the distance from the source. This arises simply from the **geometry** of the situation. The energy is spread out over the surface of a sphere; as the radius $r$ increases from R to 2R, the area increases from $4\pi R^2$ to $4\pi(2R)^2$. Thus the intensity falls to one quarter.

$$I \propto 1/r^2$$

This is the inverse square-law for radiation. It describes the attenuation of any radiation in a vacuum, and is a reasonable approximation for the attenuation of X, $\gamma$- and $\beta$-radiations in air.

In a medium, where absorption processes are occurring, the intensity of a beam $I$, falls by a constant fraction $dI/I$, through each unit distance travelled $dx$. That is

$$-dI/I = \mu dx$$

or

$$\frac{-dI}{I dx} = \mu$$

where $\mu$ is the **total linear attenuation coefficient**, a constant which depends on the medium and the photon energy of the X-rays. Integrating this differential equation gives

$$\log_e I - \log_e I_0 = -\mu x$$

where $I_0$ is the incident intensity and $I$ is the intensity at distance $x$.

$$I = I_0 e^{-\mu x} \qquad (2)$$

**FIG 9.10** Exponential absorption of radiation by a medium.

This results in an exponential fall in the intensity with distance, as shown in Figure 9.10. It is useful in radiology to define the **mass attenuation coefficient** $\mu_m$ which refers to the attenuation per unit mass of material traversed.

$$\mu_m = \mu/\rho$$

where $\rho$ is the density.

The penetrating power or quality, of a radiation can conveniently be described in terms of the thickness of material needed to reduce the intensity to half the original value. This is called the **half-value thickness**, $x_{1/2}$ (HVT). If we put $I = I_0/2$ and $x = x_{1/2}$ in equation 2, we obtain

$$x_{1/2} = \log_e 2/\mu = 0.693/\mu$$

A beam of 80 keV X-rays has an HVT of 1 mm in copper, and a cobalt-60 gamma source (1 MeV), has an HVT of 10 mm in lead. You may be able to check the latter in the laboratory, by placing various thicknesses of lead between the cobalt-60 source and a Geiger tube and counter.

## INVESTIGATION

### Half-value thickness

Measure the half-value thickness of cobalt-60 gamma radiation, for lead and for wood. Estimate the percentage reduction in gamma radiation produced by the box in which the source is supplied.

## Attenuation and photon energy

The mass attenuation coefficient of a medium is a function of its proton number Z (because of the probability of collision between photon and electron). Penetration is also dependent on photon energy of the X-rays, in different ways, depending on the mechanism of the attenuation. This is summarised in Table 9.2.

The contributions of these mechanisms to the total attenuation in the body, is shown in Figure 9.11, where the envelope curve is the sum of the attenuation of the four individual mechanisms. You will notice from this graph that as the incident photon energy is increased the radiation becomes more penetrating (the value of the mass attenuation coefficient falls). The optimal photon energy for radiography is around 30 keV, when the photoelectric effect predominates, as this gives maximum contrast between different tissues, because of its dependence on $Z^3$. For example bone, with a high atomic density, attenuates the beam about eleven times more than the surrounding tissue.

**TABLE 9.2** Attenuation mechanisms

| MECHANISM | VARIATION OF $\mu_m$ WITH $E$ (PHOTON ENERGY) | VARIATION OF $\mu_m$ WITH $Z$ | ENERGY RANGE IN WHICH IT IS THE DOMINANT MECHANISM IN SOFT TISSUE |
|---|---|---|---|
| simple scatter | $\propto 1/E$ | $\propto Z^2$ | 1 to 30 keV |
| photoelectric effect | $\propto 1/E^3$ | $\propto Z^3$ | 1 to 100 keV |
| Compton scatter | falls very gradually with $E$ | independent | 0.5 to 5 MeV |
| pair production | rises slowly with $E$ | $\propto Z^2$ | above 5 MeV |

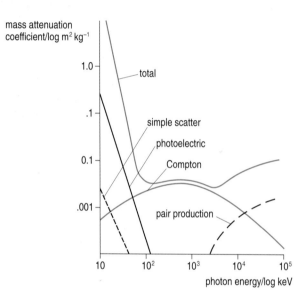

**FIG 9.11** The combined effect of the four attenuation mechanisms.

For therapeutic uses, higher energies are preferred (between 0.5 and 5 MeV), where Compton scatter predominates, as this avoids the preferential absorption of the beam by the bones, because this mechanism is independent of Z.

## Filtration

Most X-ray beams are heterogeneous when generated, containing a range of energies. This is not ideal for use so the beam is made more homogeneous by the use of selected high Z materials which filter out the lower energy photons, by the mechanism of photoelectric emission. During radiotherapy, these would only be absorbed by the patient's skin, adding unnecessarily to the dose. For the low tube voltage used in diagnosis and superficial therapy, filters usually consist of a few millimetres of aluminium. For therapeutic radiation deep in the body, higher voltages are used with filters with higher Z to reduce Compton scatter. Composite filters of copper, tin, lead and gold may be used, to select a narrow range of photon energies. The resulting hardened beam has an HVT which is greater than the original beam.

**9.5** Explain the origin of the photoelectric and Compton effects in absorbing energy. Which of these is preferred for radiography and which for therapy? Give reasons.

**9.6** Explain what kind of filter would be used with X-rays used for

(a) radiography,

(b) radiotherapy. Explain why.

**9.7** (a) An X-ray tube operates at a steady tube voltage of 70 kV and tube current of 120 mA. It produces a beam of cross-sectional area 4 mm$^2$, with an efficiency of 1 per cent. Assuming there is no inherent filtration, calculate the intensity of the beam as it emerges from the tube.

(b) Assuming that the intensity in (a) is at a distance of 0.1 m from the point source of the X-rays, calculate the intensity at a distance of 1 m from the tube.

(c) Calculate the intensity of the same source of X-rays after they have travelled through 4.5 mm aluminium, placed 1 m from the tube. The half-value thickness for the beam in aluminium is 1.5 mm.

# X-RAY EQUIPMENT

**W**e can now take a look at the stages in the creation of a radiographic image. An X-ray generator produces the radiation; apertures and grids control the geometry of the beam and so define the image; film or fluorescent screens detect the image, and various methods are used to increase its brightness; the irradiated tissue can be modified by the addition of a contrast medium to improve the image.

## The X-ray generator

Figure 9.12 shows the structure of the rotating anode X-ray tube. Tubes are evacuated glass envelopes, immersed in oil to provide cooling, and shielded by lead. The filament is a coil of tungsten wire, heated to incandescence by a low voltage supply. A metal focusing cup converges the emerging thermionic electrons to a focus on the target. The target is a solid tungsten disc, with a bevelled edge. It is rotated rapidly during operation, normally at about 3600 revolutions per minute. In making a radiograph the energy supplied can be as high as 10 kJ in 0.2 s but only about 1 per cent of this is converted into X-rays, so the rest has to be dissipated. Tubes used for therapy need to supply more total energy, but over a longer period; typically 2.25 MJ over 600 s. For these the tungsten target is fixed and mounted on a copper anode block through which oil is pumped.

The X-ray tubes operate at potential differences of up to 40 MV, and at tube currents in the range of tens of mA. These are supplied by high voltage transformers, with a low voltage transformer for the heater current to the filament.

**FIG 9.12** A rotating anode X-ray tube.

**9.8** How is the energy dissipated in the two types of tube described above? Why do you think different methods of cooling are used in the two types of X-ray tubes?

## Controlling the X-ray beam

The ideal beam is small and sharp edged, so that it will produce clear images in radiography and can deliver all its energy to the selected tissue in radiotherapy. The beam of electrons is focused on to the target in the X-ray tube, and the X-rays then radiate from this spot. Since they cannot be focused, X-ray images are basically shadows of objects placed in the beam. You can investigate the process of improving the sharpness of a shadow by placing your hand in the beam of a lamp.

**9.9** Why can X-rays not be focused like electrons?

### Shadows

If you take an ordinary reading lamp and the bulb is large the shadow of your hand will be blurred.

1. What happens to the shadow when the lamp is small or a pinhole is interposed as in Fig 9.13(a)?

**2.** The sharpness can also be improved by changing the relative positions of hand, screen and lamp. Try this. The region of blur or partial shadow is called the **penumbra**. Its dependence on separation can be seen from the geometry of similar triangles – Fig 9.13(b).

**3.** The material surrounding the object will also affect its image. If you place your hand in cloudy water which is in the beam, some of the light is scattered and some absorbed, so the shadow loses sharpness and brightness – Fig 9.13(c).

**4.** It will be clear to you without a demonstration, that any movement of your hand will also reduce the sharpness of the image.

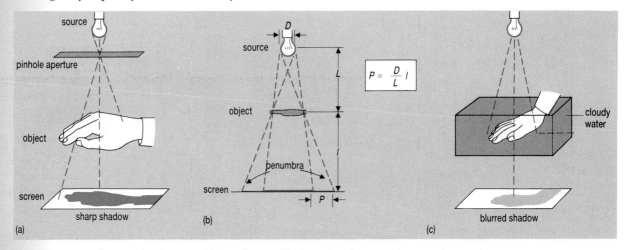

**FIG 9.13** Factors affecting image formation **(a)** size of source **(b)** relationship of position of source, object and screen **(c)** medium surrounding the object.

**QUESTION**

**9.10** Relate these observations to the formation of an X-ray image by the lungs of a patient.

We will now take a look at the ways in which the beam has to be limited physically. Orienting the target at an acute angle (typically 17°), to the direction of the electron beam, spreads the area over which the X-rays are generated. This minimises the temperature rise while reducing the apparent area of the X-ray beam. This **line-focus** principle, illustrated in Figure 9.14, is simply a foreshortening of the line (actually an area in three dimensions), by the angle of view.

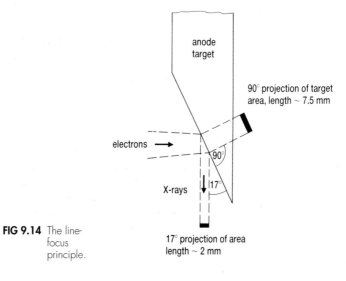

**FIG 9.14** The line-focus principle.

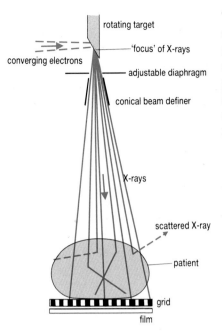

**FIG 9.15** Directing the X-ray beam.

The beam can then be limited or defined, by the use of apertures. Two types are shown in Figure 9.15, the adjustable diaphragm and the simple cone. They are made of lead to absorb scattered radiation. Scattering will occur in the tissue being examined, this can cause blurring of the image. It can be reduced by the use of a grid, directly in front of the detector. It consists of strips of lead, 0.05 mm thick and 5 mm long, sandwiched between material which is transparent to X-rays. This arrangement allows direct radiation to fall on the screen, and absorbs scattered radiation.

## Creating the image

Since the time of Roentgen, the photographic film and the fluorescent screen have been used to display X-ray images. The film is used when a record of the image is required. The use of screens, called **fluoroscopy**, enables real-time investigations to be carried out. If for example a radiologist is examining the bowel, she can move this around whilst viewing (with her hand suitably protected with a lead glove!), to see all the details. In practice the use of these screens requires a high dose of X-rays to produce an image bright enough to view directly. Those used in shops as a gimmick to sell shoes were banned as they gave an unacceptably high dose, especially to the testes of small boys. The dose can be reduced by means of an **intensifying screen** which usually contains zinc sulphide as the fluorescent material. This absorbs the X-radiation and re-emits it in the visible region. The structure of the screen is shown in Figure 9.16, with the double sided film placed to receive X-radiation direct from the patient, and light from both the screens. In practice these can intensify the image by up to 40 times, although there is some loss of definition, due to the spreading out of the fluorescent light. The resolution of direct exposure film is about 0.1 mm, this can fall to 1 mm with fluoroscopic screens. This is used for still photography, to produce radiographs. For direct observation or recording of movement, light of a higher intensity is required.

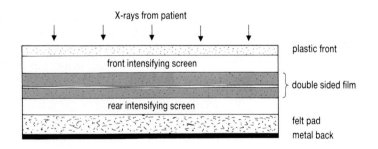

**FIG 9.16** An image intensifying screen.

The use of the **image intensifier tube** overcomes the need for a greater intensity of X-rays by coupling the fluorescent screen to a photocathode and electron gun (Figure 9.17). The electrons emitted by the photocathode are accelerated on to a second fluorescent screen containing zinc sulphide, the energy they have gained intensifying the original image, usually by a factor of up to 1000. The signal can then be fed to a video camera for 'real-time' viewing using a television monitor or video recording.

**FIG 9.17** An image intensifier tube.

## 9.4 THE USE OF X-RAYS IN DIAGNOSIS

**X**-ray images or radiographs of the human body are clearest when there is high **natural contrast** between the tissues it is required to distinguish. This occurs when there is the greatest difference in the proton number, Z, of the tissues, because the photoelectric effect depends on Z. The values of Z are high for bone, moderate for soft tissue and low for air. This is the reason why radiographs of the **skeleton** are used to diagnose fractures, dislocations and other abnormalities, such as arthritis (Figure 9.18).

**FIG 9.18** X-radiograph of skeleton showing arthritis.

For soft tissue imaging contrast can be improved by the introduction of an **artificial contrast medium.** For example, in investigations of the gastrointestinal tract, which is all soft tissue, the patient is given a 'barium meal' to swallow or a barium enema. Barium has a high Z so shows as an opaque area in the image (Figure 9.19).

In an area of continuous soft tissue such as the breast, careful choice of the X-ray beam and film detector is used to maximise contrast, in screening for tumours. Low tube voltages are used with molybdenum targets, so the photoelectric attenuation predominates. Special mammographic film cassettes and a small focal spot are used to provide a high resolution. The image is also improved by using relatively high tube current with a short exposure time to minimise movement blur.

**Computed tomography** is a good method for examining soft tissue, as it can detect small differences in attenuation of X-rays. The tomographic technique removes the interference from other structures which are out of the chosen plane. Tomography is explained in Chapter 12.

**FIG 9.19** Barium enema in the large bowel, showing as white contrast. The darker parts of the bowel are filled with air.

## 9.5 RADIOTHERAPY

The purpose of radiotherapy is to destroy malignant cancer cells in the body, without damaging normal cells or interfering with the unaffected part of the body. It may seem paradoxical that radiations can cure the very diseases that they can cause. It turns out that all cells are susceptible to radiation damage but those which are malignant are, under certain circumstances, more susceptible than normal cells. In Chapter 11 you will learn more of this. At this stage we will concentrate on descriptions of the techniques and equipment used.

### *Energy delivery*

The most important feature of therapy is delivering the energy in the correct amount, to the affected tissue and not to other parts of the body. This is increasingly difficult the further into the body the malignant tumour is situated. There is a greater likelihood of damaging the intervening and surrounding tissue. The actual dose received by the tumour is critical. Typical results indicate that if a particular dose can cure, then 10 per cent less can leave the tumour unaffected and 10 per cent more can cause other problems.

Radioisotope tracers offer a completely different approach; substances are taken into the body so that they accumulate only at the sites required, to deliver the radiation energy locally. These methods are described in the next chapter. Here we are concerned with the longer established, and still more widespread, use of external sources.

### *X-ray equipment*

X-rays for diagnostic use are in the kilovoltage (kV) range. This produces good contrast as X-rays are differently absorbed. In order to destroy tissue, X-rays in the **megavoltage (MV)** range must be used, for two reasons.

■ At these voltages the absorption of energy is mainly by Compton scatter which is not dependent on the tissue type or proton number, rather than by the photoelectric effect which is (see Table 9.2). This means that energy can be dissipated in whatever tissue is required.

- Higher energy beams are more penetrating, and so do not damage the skin or intervening tissue as much. A rough guide is that the penetration increases by 2.5 mm for each increase in voltage of 1 MV.

The development of X-ray machines for therapeutic use has therefore been concerned with producing higher energies accurately and reproducibly. Although the therapeutic use has a long history, the highest photon energy levels possible with conventional X-ray tubes are of the order of 0.1 MeV. The possibilities were extended by the invention of the Betatron about fifty years ago. This can accelerate electrons to voltages of up to 40 MeV producing very penetrating photons of energies up to 10 MeV, but the equipment is very cumbersome. The last thirty years have seen the development of the medical linear accelerator, or **linac**, which is much more compact.

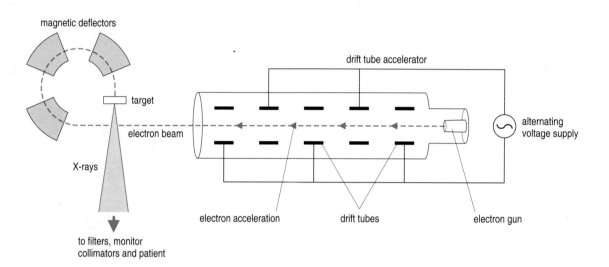

**FIG 9.20** X-ray generator for linac.

## Linear accelerator (linac)

Electrons are produced thermionically, as in a conventional X-ray machine. The key to the higher energy is the acceleration of these to near the speed of light in an evacuated tube. This requires a very high voltage across the ends. To minimise insulation problems, the tube may be spilt into sections and an alternating voltage applied across each (Figure 9.20).

There is no field inside these sections, called drift tubes, electrons are accelerated between them. The electrons are then magnetically deflected to hit the target on the far side from the patient so that the X-rays are transmitted through the target.

The generator is contained in a gantry, mounted above the treatment couch. The details of the treatment are fed into a computer to control all the movements and emissions of the gantry. All parts of the equipment can be rotated so that the tumour is the **isocentre**, always receiving the maximum dose while the dose elsewhere is much lower.

The nature of the beam transmitted to the patient is filtered and collimated, in a similar way to a conventional X-ray machine. The shape of the 'treatment field' for the patient can be modified using wedges or compensators. They are usually made of lead and are designed to protect particular parts of the body which will be exposed, and to allow for the irregular shape of the body.

**Alignment** of the beam is achieved by the use of an optical system which may include a video or laser beam. Before treatment begins an X-ray picture can be taken using lower energy X-rays, to check that everything is set up correctly. The treatment head also includes a dose monitor.

## Gamma ray therapy

Powerful artificially radioactive substances can be produced in nuclear reactors. One of these is an isotope of cobalt called cobalt-60 (details of the production of isotopes and their naming are given in the next chapter). This emits penetrating gamma rays of an energy of about 1.25 MeV which is equivalent to X-rays generated at 3 MV. The source is very small, though when suitable shielding and other equipment is added the unit is similar in size to the linac (Figure 9.21). The equipment requirements are much simpler than for X-ray therapy, and so the equipment is cheaper to maintain. There are no high voltage hazards but the cobalt source emits constant radioactivity. This cannot be turned off like X-rays; cobalt has a half-life of 5.3 years. In use it is surrounded by thick lead shielding, with a shutter which opens during the therapy period. The equipment has other disadvantages; the beam definition is poorer and the time for treatment is longer. This method tends to be replaced by the linac in many situations except where running costs are the critical factor.

**FIG 9.21** A cobalt-60 teletherapy machine.

A recent development in radiation therapy is the Gamma knife. This uses the power of modern computing to accurately position beams from several hundred cobalt-60 gamma sources. It has been developed as a neurosurgical tool for treating brain tumours and other disorders such as pain and movement problems.

A summary of therapeutic radiations is given in Table 9.3. The energy of the cobalt-60 and other gamma therapy sources is in the megavoltage range, the only difference is in the method of production. Other similarities between X-rays and radioisotopes will be considered in the following chapters.

**TABLE 9.3** Therapeutic radiations

| TREATMENT | VOLTAGE RANGE | TYPICAL PHOTON ENERGY MeV |
|---|---|---|
| Superficial therapy | 20–150 kV | 0.03 |
| Orthovoltage therapy | 140–300 kV | 0.1 |
| Mega- or supervoltage therapy | | |
| Linear accelerator | 4–8 MV | 2 |
| Betatron | 9–42 MV | 8 |
| $\gamma$ Teletherapy | | |
| $^{137}$Cs | | 0.66 |
| $^{60}$Co | | 1.17, 1.33 |

**9.11** The main advantages of megavoltage therapy compared to orthovoltage are:
**(i)** ability to treat tumours deep inside the body;
**(ii)** reduction of the dose to the skin, which considerably reduces the pain from the treatment.
**(iii)** reduction of the dose to the bone which reduces the possibility of damage.
Explain these features, with reference to the interaction of X-rays of different energies with tissue.

## SUMMARY ASSIGNMENTS

**9.12** Write an illustrated account of how the following factors affect the quality of an X-ray image of body tissue:

tube voltage, tube current, focal spot size, beam definers, filtration, intensifying screens, contrast media and exposure time.

**9.13** Sketch a graph to show how the intensity of the X-rays from a tungsten target in an X-ray tube varies with photon energy.

What effect would an increase in the X-ray tube voltage have on this graph?

The intensity of an X-ray beam passing through matter is reduced by absorption and by scattering. Explain the difference between *absorption* and *scattering*. Name and describe briefly two interactions which involve either one or both of these processes.

Explain the term *half-value thickness* applied to a tissue irradiated with a monochromatic X-ray beam.

Why is it best to use high energy X-ray photons for therapy?

*(London June 1998)*

**9.14** Write an essay about the mechanisms by which X-rays are attenuated in matter. Your account should include
**(a)** a description of the photoelectric effect, Compton scattering and pair production,
**(b)** a description of the dependence of each mechanism on photon energy and the nature of the target nuclei,
**(c)** a discussion of the advantages of the use of the photoelectric effect over Compton scattering for imaging body structures.

*(OCR June 1999)*

**9.15 (a)** State what is meant by the *mass attenuation coefficient*, $\mu_m$, and explain how it is different from, and therefore more useful than, the *linear coefficient* $\mu$.
**(b)** Explain the difference between simple or elastic scattering and Compton scattering of photons.
**(c)**

| absorption mechanism | variation of $\mu_m$ with Z (atomic number) | photon energy range in which mechanism is significant or dominant |
|---|---|---|
| simple or elastic scattering | $\propto Z^2$ | 1 to 20 keV |
| photoelectric effect | $\propto Z^3$ | 20 to 50 keV |
| Compton scattering | independent of Z | 30 keV to 10 MeV |

With the help of the table explain why 25 keV X-rays are suitable for diagnostic applications.

*(NEAB June 1998)*

**9.16** The diagram (Figure 9.22) shows the design of an X-ray image intensifier.

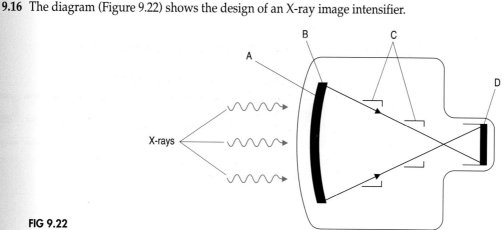

**FIG 9.22**

(a) The main components are labelled A to D. Name each component and state its purpose in the process of image intensification.

(b) Give **one** example of a medical application for which an image intensifier might be used and explain why the use of the image intensifier is required.

*(NEAB Mar 1998)*

**9.17** Give an account of the attenuation of X-rays and their use in therapy. Your account should include

(a) an explanation of the mechanisms of attenuation of X-rays in matter,

(b) a description of the relative importance of each mechanism with respect to photon energy and mass number of the target material,

(c) an outline of methods employed to reduce damage to intervening healthy tissue during therapy of a deep-seated tumour.

*(UCLES Feb 1997)*

**9.18** X-rays used for diagnosis are generated with tube voltages of around 70 kV compared with several megavolts when the X-rays are needed for therapy. What is the difference between diagnosis and therapy?

Explain why 70kV is enough for diagnosis but up to several megavolts is often needed for therapy.

*(London June 1997)*

# 10 RADIONUCLIDES

**T**hree types of radiation are emitted spontaneously from the nuclei of certain atoms. Gamma radiation is similar to X-rays, but beta and alpha radiations are substantially different in their nature and in the way they interact with matter. In this chapter we will study how this results in a wide range of diagnostic and therapeutic applications of these radiations.

### LEARNING OBJECTIVES

*After studying this chapter, you should be able to:*

① define and use the following terms: radioactive decay constant, half-life, biological half-life, activity, specific activity, meta-stable radionuclide, tracer;

② state the properties of alpha, beta and gamma radiations and the ideal radiopharmaceutical;

③ give an example of the production of radionuclides in a nuclear reactor and in a cyclotron;

④ describe the production and use of technetium-131 and iodine-131 as tracers;

⑤ give examples of other uses of radionuclides for diagnosis and therapy, explaining their suitability for the function.

## 10.1 RADIOACTIVE DECAY

**A**n atom of a particular element X, is represented symbolically as $^A_Z X$, where

$A$ is the **nucleon number**, or mass number
$Z$ is the **proton number**, or atomic number.

Examples are $^1_1 H$, $^{16}_8 O$, and $^{206}_{82} Pb$, for hydrogen, oxygen and lead. Each variety of nucleus is called a **nuclide**, and has a specific value of $A$ and $Z$. Nuclides with the same proton number (and therefore the same chemical nature) but different nucleon numbers, are called **isotopes**. Most elements have naturally occurring isotopes, but when the ratio of proton to neutrons falls outside a certain range, the nuclide is unstable. It decays to another nuclide, with the emission of radiation. Isotopes of hydrogen are $^2_1$H (deuterium – stable), and $^3_1$H (tritium – unstable, $\beta^-$ emission). Lead has six isotopes, of which three are stable and three radioactive, one of which is $^{214}_{82}$Pb. Alternatively this can be referred to as $^{214}$Pb, Pb-214, or lead-214. Often there is a series of decays until a stable nuclide is reached. This is natural radioactivity. Radioactivity can also be induced by bombarding stable nuclei with nuclear particles in a nuclear reactor or cyclotron.

## Decay law

The process of decay is entirely random, depending only on the nature of the nuclide. The rate of emission of radiation can be described statistically, due to the large number of atoms present, even in a small mass of substance. The number of atoms that decay in unit time is directly proportional to the number of atoms present in the sample, that is:

$$-dN/dT \propto N$$
$$\text{or} \quad -dN/dT = \lambda N$$

where $N$ is the number of atoms remaining and $\lambda$ is the **radioactive decay constant**. Suppose the number of active nuclei at time $t = 0$ is $N_0$ and after time $t$ this is $N$. Then integrating over time $t$ gives:

$$\int_{N_0}^{N} \frac{dN}{N} = -\lambda \int_0^t dt$$

$$\log_e \frac{N}{N_0} = -\lambda t \quad \text{or} \quad N = N_0 e^{-\lambda t}$$

(1)

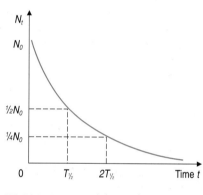

**FIG 10.1** Exponential decay of a radioactive nuclide.

The number of parent nuclei present decays **exponentially**, as shown in Figure 10.1. (Compare this with the attenuation of X-rays with distance, Figure 9.10 and equation 2 of Chapter 9.)

## Activity

The rate of disintegration of a radioactive nuclide is its activity $A$, which is defined by;

$$A = dN/dt$$

So the unit of activity is second$^{-1}$, which is called the **becquerel**, symbol Bq. The activity is also given by $A = \lambda N$, and the change in activity from $A_0$ at time $t_0$, to $A_t$ at time $t$ can be written by substituting in equation 1, as :

$$A_t = A_0 e^{-\lambda t}$$

showing that the activity also falls exponentially. The term $e^{-\lambda t}$ is called the **decay factor**; it is the fraction of the initial activity present after time $t$.

The **specific activity** is an important quantity in the handling and preparation of radionuclides. It is defined as the activity of a sample divided by its total mass, $A/m$, and is measured in becquerels per kilogram, Bq kg$^{-1}$. Samples will have varying amounts of active nuclei, depending on their chemical nature, whether there is a mixture of isotopes present and how much decay has occurred. For example, naturally occurring uranium is mainly the isotope U-238. For use in a nuclear reactor it is 'enriched' to increase the proportion of the isotope U-235 with a higher activity. A sample of enriched uranium thus has a higher specific activity.

## Half-life

The half-life of a radionuclide $T_{1/2}$, is the time taken for half of the active nuclei to disintegrate, or alternatively the time taken for the activity to fall to half its original value (Figure 10.1).

Since $\qquad N_t = N_0 e^{-\lambda t}$

$\qquad\qquad \frac{1}{2}N_0 = N_0 e^{-\lambda T}$

therefore $\quad \lambda T_{1/2} = \log_e 2 = 0.693$

and $\qquad T_{1/2} = 0.693/\lambda$

The half-lives of a number of important radionuclides are shown in Table 10.1.

When a radioisotope is taken into the body it is subject to various biological processes which will remove it, such as respiration, urination and defecation. This means that its effective half-life $T_E$ in the body is less than the physical half-life $T_R$ from radioactive decay alone. The **biological half-life** $T_B$ is defined as the time taken for biological processes to remove half the original active material. The effective half-life is then given by the equation:

$$\frac{1}{T_E} = \frac{1}{T_B} + \frac{1}{T_R}$$

**TABLE 10.1** The half-lives of some common radioisotopes

| RADIOISOTOPE | HALF-LIFE |
|---|---|
| *Natural* | |
| hydrogen–3 (tritium or triton) | 12.3 a (years) |
| carbon–14 | 5730 a |
| potassium–40 | $1.3 \times 10^9$ a |
| radon–220 | 55.5 s |
| radium–226 | 1622 a |
| uranium–238 | $4.51 \times 10^9$ a |
| *Artificial* | |
| sodium–24 | 15 hours |
| phosphorus–32 | 14.3 days |
| cobalt–60 | 5.26 a |
| iodine–125 | 60 days |
| iodine–131 | 8 days |
| caesium–137 | 30 a |
| thallium–208 | 3.1 minutes |

For example, iodine-131 is used to label the human serum albumen. Iodine-131 has a physical half-life of 8 days, but is removed from the body with a half-life of 21 days. The effective half-life can be calculated as

$$\frac{1}{T_E} = \frac{1}{21} + \frac{1}{8}$$

so $\quad T_E = 5.8$ days.

The biological half-life depends on metabolic rates which vary between different organs of the body and between individuals. It is difficult therefore to give accurate values and so to calculate accurate figures for the effective half-life which would enable the radioactive dose to be known.

QUESTIONS

**10.1** The body's ability to metabolise radionuclides can be used to reduce the dose by substituting the absorbed radioactive nuclide with a non-radioactive one. One of the principal isotopes in the fallout from the Chernobyl nuclear power station accident, was iodine-131. Find out what treatment was given to those affected, to reduce their exposure.

**FIG 10.2** A Polish child being dosed after Chernobyl.

**10.2** This question requires a calculation on half-life, activity and specific activity. Naturally occurring uranium contains 99.28 per cent U-238 with a half-life of $4.51 \times 10^9$ years, and 0.72 per cent U-235 with a half-life of $7.13 \times 10^8$ years. The specific activity can be calculated as follows:

Activity $A = \lambda N \log_e 2 / T_{1/2}$

One mole of uranium has a mass of 0.238 kg, and contains $6.022 \times 10^{23}$ atoms of uranium (the Avogadro constant). The activity of the U-238 is therefore

$$\frac{0.9928 \times 6.022 \times 10^{23} \times 0.693 \text{ Bq}}{4.51 \times 10^9 \times 365 \times 24 \times 60^2}$$

which is a specific activity of $2\,913\,000/0.238$ Bq per kilogram
$= 12.24$ MBq kg$^{-1}$

The activity of the U-235 is

$$\frac{0.0072 \times 6.022 \times 10^{23} \times 0.693 \text{ Bq}}{7.13 \times 10^8 \times 365 \times 24 \times 60^2}$$

which is a specific activity of $1\,36\,890/0.235$ Bq per kilogram
$= 0.57$ MBq kg$^{-1}$

So the total specific activity of the natural uranium is 12.81 MBq kg$^{-1}$
Uranium for use in a nuclear reactor is enriched by replacing some of the U-238 with U-235, so that 'enriched' uranium contains 3.2 per cent U-235. What is the specific activity of this?

# THE NATURE AND PROPERTIES OF
## NUCLEAR RADIATIONS

**10.2**

The alpha, beta and gamma radiations all originate from the atomic nucleus. Their emission changes the nature of the nuclide in different ways, and the behaviour of each type of radiation is characteristic. This was discovered by Rutherford and co-workers in the early years of the twentieth century, and the work is described in the physics course textbook and the topic book on nuclear physics. Here we will summarise the results and look in detail at the effect of the radiations on the body.

## *Emission reactions*

An $\alpha$ particle is emitted in the decay of many elements with a proton number greater than lead (i.e. $Z > 82$). It is identical to a helium nucleus, consisting of two protons and two neutrons. It therefore has a charge of $+2e$ and a rest mass of about $7 \times 10^{-27}$ kg. Emission results in a reduction of the proton number, $Z$, by 2 and in the nucleon number, $A$, by 4. An example is radium-226, the radioisotope used in a school cloud chamber, which decays to radon:

$$^{226}_{88}\text{Ra} \rightarrow \, ^{222}_{86}\text{Rn} + \, ^4_2\alpha$$

### QUESTION

**10.3**    Another isotope of radon, Rn-220 is the product of the decay of thorium. It decays by $\alpha$ emission with a half-life of 55.5 s, so it can be readily observed in cloud chambers. Write down the equation for Rn-220 decay.

A $\beta^-$ particle is emitted by elements having an excess of neutrons over protons. It is simply an electron which is emitted by the decay of a neutron into a proton:

$$^1_0\text{n} \rightarrow \, ^1_1\text{p} + \beta^-(\text{e}^-) + \upsilon$$

$\upsilon$ represents an antineutrino. This is the antiparticle to the neutrino and it carries away some of the energy of the disintegration. $\beta^-$ emission results in no change to the $A$ of the nuclide, and an increase in the $Z$ by 1. An example is the school laboratory source strontium-90, which decays to produce yttrium:

$$^{90}_{38}\text{Sr} \rightarrow \, ^{90}_{39}\text{Y} + \text{e}^- + \upsilon$$

A $\beta^+$ particle or positron is emitted from elements which have a deficit of neutrons compared to protons. It is simply a positively charged electron which is emitted by the decay of a proton into a neutron:

$$^1_1\text{p} \rightarrow \, ^1_0\text{n} + \beta^+(\text{e}^+) + \upsilon$$

$\upsilon$ represents the neutrino. $\beta^+$ emission results in no change to the $A$ of the nuclide, and a decrease in the $Z$ by 1. Take for example the isotopes of oxygen, proton number 8. The isotopes with 5, 6, and 7 neutrons decay by $\beta^+$, those with 8, 9, and 10 are stable, and those with 11 and 12 neutrons decay by $\beta^-$.

The neutrino is a massless, uncharged particle which carries away some of the energy in $\beta^+$ decay. The antineutrino is its counterpart in $\beta^-$ decay. As it has no mass or charge it is difficult to detect. It was only discovered many years after the quantum theory predicted its existence, because of the need to have a particle of this amount of energy.

QUESTION

**10.4**  Write down the nuclear equation for the β decays of **(a)** O-14 and **(b)** O-19.

**Gamma rays** are emitted from the nucleus following β or α emission. The daughter nuclide is left in an excited state by the decay; γ emission enables it to return to a low energy state. Usually this occurs immediately after the main decay, sometimes the daughter can exist in a metastable state for some time, delaying the γ emission. A source of pure γ radiation can be produced by creating such a metastable radioisotope, an example is the physics laboratory source cobalt-60. Gamma emission does not change either the *A* or the *Z* of the nuclide.

**Electron capture** is a type of nuclear change in which a proton combines with an electron from an inner shell, usually the K-shell, to produce a neutron. This has the same effect on the nucleus as β⁺ decay, leaving *A* unchanged and reducing *Z* by 1. An example is the decay of chromium into vanadium:

$$^{51}_{24}\text{Cr} + e^- \rightarrow {}^{51}_{23}\text{V} + \upsilon$$

Subsequently an electron will move from an outer orbit to fill the space left by the captured K-electron, and a characteristic X-ray photon is emitted.

## Properties of nuclear radiations

Table 10.2 summarises many of the important properties of the radiation.

The decay of a particular nuclide usually results in emissions of α, β and γ with a spectrum of energies; continuous in the case of β and with characteristic energies for α and γ. This will affect the radiation's subsequent ionisation of the material into which it passes and so also its penetration. Alpha particles ionise any medium rapidly and give up their energy over a short path distance; they are said to be high linear energy transfer (LET) radiation. Beta interacts in a similar way but less frequently, and the particle is readily scattered in the interaction. Gamma rays ionise only occasionally, by three mechanisms which are the same as those for X-rays, described in Section 9.2.

**TABLE 10.2** Comparative properties of nuclear radiations

|  | α | β | γ |
|---|---|---|---|
| Nature | helium nucleus | electron ß⁻ or positron ß⁺ | high frequency photon or e-radiation |
| Charge | +2e | −e or +e | none |
| $\dfrac{\text{Mass}}{\text{Mass of proton}}$ | 4 | 1/2000 | none |
| Energy range from a particular nuclide | spectrum with characteristic emissions | continuous spectrum | spectrum with characteristic emissions |
| Ionisation mechanisms | frequent collision with orbital electrons | ejection of electrons, some secondary ionisations | low energy − photoelectric effect medium energy − Compton effect high energy − pair producton |
| Number of pairs produced per mm of air | 1000 | 10 | 1 |
| Radiation absorbed by | 10⁻² mm Al or a few cm air | 1 mm Al | 10 cm Pb |

## QUESTION

**10.5** Use the information in Table 10.2 to answer the following:

    **(a)** How would you expect the deflection of the three radiations, by electrical and magnetic fields, to compare?

    **(b)** Why does γ radiation have a fixed speed? What is changed in γ rays of different energies?

    **(c)** Fig 10.3 shows the tracks that can be observed when radium is placed in the chamber. Explain fully the differences between the three types of track. If you have not made observations like these, you should see if you can set up a cloud chamber now.

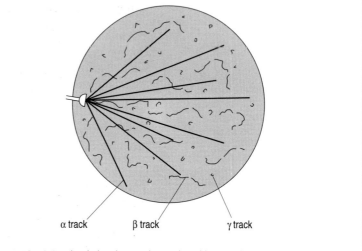

α track      β track      γ track

**FIG 10.3** Cloud chamber tracks produced by a radium source.

# Penetration of tissue

The overall process of interaction of nuclear radiations with tissue is very similar to that described in Section 9.2. The absorption of energy causes ionisation and excitation, which results in chemical changes, with a range of biological effects. The absorption of the energy in tissue is quite different, between each type of radiation however, as you would expect from studying Table 10.2. This is summarised in Figure 10.4.

**Alpha particles** have very little penetration into any type of tissue. An external source will be stopped by the air, unless closer than a few centimetres, when it will be absorbed by a few mm of skin. A source which is ingested however, will be very damaging, because of the density of the internal ionisation.

**Beta particles** have a wide range of penetrations and because of their irregular paths, their depth in tissue cannot be accurately defined. A maximum depth can be estimated and is considerably greater than α, with the ionisation damage being spread throughout this depth to a decreasing extent as shown in Figure 10.4(b).

**Gamma rays** do not have a definite penetration depth in tissue, because their interaction causes them, like β to follow irregular paths, and their intensity shows an exponential attenuation like X-rays. The intensity $I$ at thickness $x$ is given by:

$$I = I_0 e^{-\mu x}$$

(This is strictly true only for a point source in a vacuum.)

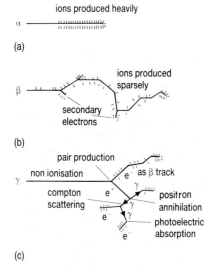

**(a)** α — ions produced heavily

**(b)** β — ions produced sparsely — secondary electrons

**(c)** γ — non ionisation — pair production — compton scattering — $e^-$ as β track — positron annihilation — photoelectric absorption

**FIG 10.4** Interaction of radiation with tissue **(a)** alpha **(b)** beta **(c)** gamma.

where $I_0$ is the incident intensity and $\mu$ is the linear attenuation coefficient of the medium, (see Section 9.2). The penetration is described by the **half-value thickness**, for cobalt-60 the HVT with lead is 12.5 mm. It is not possible to screen from $\gamma$ completely in most situations. However the damage caused is considerably less than $\alpha$ and $\beta$ because of the low ionisation rate. Remember also that $\gamma$ rays, like other electromagnetic radiations, attenuate in free space (or air), according to the inverse square law:

$$I = I_0 r_0^2 / r^2$$

---

### QUESTION

**10.6**   A cobalt-60 source gives a dose rate at 1 m from it, of 80 µSv h$^{-1}$, (Microsievert per hour; you will learn how the sievert is defined in the next chapter).
   **(a)**  At what distance from the source is the dose rate 25 µSv h$^{-1}$?
   **(b)**  What thickness of lead, placed 1 m from the source would give the same protection?

---

## Neutron radiation

Neutrons can be emitted spontaneously, for example in the fission of uranium-235, or emitted as a result of the bombardment of nuclei with energetic particles. These neutrons, because they are uncharged, can penetrate into target atoms and interact with the nuclei. Neutrons with energies of about 1 MeV are called fast neutrons. They can collide with several nuclei and cause considerable ionisation before losing their energy. Neutrons with energies in the eV range, called slow or thermal neutrons, are usually captured by nuclei, which then emit $\gamma$ radiation. Neutrons have an approximately exponential attenuation through tissue, like $\gamma$ rays. They are absorbed particularly well by hydrogen atoms, which are in plentiful supply in water in the body.

## 10.3   THE PRODUCTION OF RADIONUCLIDES

**M**ore than 2000 radionuclides have been made artificially, in addition to the 200 or so naturally occurring ones. The only naturally occurring radioisotope still used is radium-226 (first used in 1901) and this is now generally being replaced by others. Carbon-14 is also used in some investigations, though the source is artificially produced. There are two methods of production, by bombarding stable nuclei with neutrons in a nuclear reactor or with charged particles in an accelerator such as a cyclotron. Not all of these 2000 are potentially useful; there are many considerations to determine an ideal radionuclide. One of the most significant is whether it is to be used for diagnosis or for therapy.

In addition to the particular species of nuclide which is to be used, the specific activity is an important factor, and the pharmaceutical form in which the isotope is prepared. In this section we shall see that different methods are appropriate for certain radionuclides, and illustrate this with reference to technetium-99 and iodine-131.

# Nuclear reactor production methods

Fission reactors operate with natural uranium which absorbs neutrons to form the unstable uranium-235. This splits to form two approximately equal fragments, neutrons and the release of energy:

$$^{235}_{92}U + ^1_0n \rightarrow ^{236}_{92}U \rightarrow ^{138}_{53}I + ^{95}_{39}Y + 3^1_0n + \text{energy}$$

This results in a large number of useful isotopes as fission products, for example iodine, strontium and caesium. Although it is quite possible to extract these chemically from the fuel rods of the reactor where they are formed, this is expensive, because of the high radiation. A more common use of the reactor is to irradiate samples or targets with the excess neutrons produced by fission. The neutrons are absorbed by stable nuclei, which then typically become $\beta^-$ emitters.

---

**QUESTION**

**10.7** Refer back to Section 10.2 to explain why neutron irradiation produces $\beta^-$ emitters.

---

There are three main types of reaction.

**(n, $\gamma$) reactions** or neutron captive reactions are the most frequently used type. The irradiated nuclide captures a neutron and emits a $\gamma$ ray. For example:

$$^{31}_{15}P + ^1_0n \rightarrow ^{32}_{15}P + \gamma$$

which is commonly abbreviated to:

$$^{31}P(n,\gamma)^{32}P$$

---

**QUESTION**

**10.8** Other target materials used include $^{23}$Na, $^{41}$K, and $^{59}$Co, the latter to produce the familiar $^{60}$Co source. Write the equations for their production, and hence the general equation for a target of $^A_Z$X.

---

These products are chemically identical with their target materials, so separation is not possible. Since only a small fraction of the sample undergoes neutron capture, there will be a high proportion of carriers in radionuclides produced this way and it will have a low specific activity. In some cases the product undergoes $\beta^-$ decay to produce a useful nuclide which can then be chemically extracted. The widely used iodine-131 is produced from tellurium this way, for example:

$$^{130}_{52}Te(n,\gamma)^{131}_{52}Te(\beta^-)^{131}_{53}I$$

**(n, p) reactions** are less common and require higher neutron energies than the previous reaction. A proton is ejected from the nucleus as a result of the absorption of the neutron. This changes Z so the product can be chemically separated from the target. An important example of this is the production of carbon-14 from nitrogen, for use as a labelling tracer of organic molecules in biological and medical research.

$$^{14}N(n,p)^{14}C$$

**(n, $\alpha$) reactions** also require fast neutrons, to remove an $\alpha$ particle when the neutron is captured. The only important medical example is the production of tritium, an isotope of hydrogen, from lithium.

$^6\text{Li}(n,\alpha)^3\text{H}$

---

### QUESTION

**10.9** Write the two previous nuclear reactions in the fuller form, using the $^A_Z X$ notation.

---

## Cyclotron production methods

In order to penetrate the nucleus with charged particles, energies of several MeV must be supplied, for which a cyclotron is usually used. The bombarding particles are usually deuterons, or $\alpha$ particles, though protons are sometimes used. The product is always chemically different, and its decay is by $\beta^+$ or electron capture. Examples of reactions are:

**(d,n)** Oxygen-15 is produced by irradiating nitrogen-14 with deuterons:

$$^{14}_{7}\text{N} + {}^{2}_{1}\text{H} \rightarrow {}^{15}_{8}\text{O} + {}^{1}_{0}\text{n} \quad \text{or} \quad {}^{14}\text{N}(d,n)^{15}\text{O}$$

**($\alpha$,np)** Fluorine-18 is produced from oxygen irradiated with $\alpha$ particles:

$$^{16}_{8}\text{O} + {}^{4}_{2}\text{He} \rightarrow {}^{18}_{9}\text{F} + {}^{1}_{1}\text{p} + {}^{1}_{0}\text{n} \quad \text{or} \quad {}^{16}\text{O}(\alpha,pn)^{18}\text{F}$$

**($\alpha$,2n)** Iodine-123 is produced by irradiating antimony-121 with $\alpha$ particles. This isotope is preferred to iodine-131 because it emits only $\gamma$ rays.

$$^{121}\text{Sb}(\alpha,2n)^{123}\text{I}$$

---

### QUESTION

**10.10** Complete Table 10.3 which summarises the production of radionuclides by the two methods.

---

**TABLE 10.3** Summary of the production of radioisotopes

|  | NUCLEAR REACTOR | CYCLOTRON |
|---|---|---|
| Mode of operation | ............ bombardment and fission | Bombardment by protons, deuterons and ............ particles |
| Major reactions | (n,γ), (n,p) (n,α) U (n,f) | (d,n), (α,d) (α,np) (p,n) and others |
| Neutron-proton ratio | ................................. | ................................. |
| Mode of decay | ............ | ............, EC |
| If carrier free | (n,γ) ......................... (n,p) (n,α) } Yes U (n,f) | ................................. |
| Cost | Comparatively low | Comparatively high |

**FIG 10.5** Hospital cyclotron at the Texas Medical Centre, Houston, USA.

The majority of radionuclides are produced in reactors. Cyclotrons are a more expensive way of producing radionuclides because only small quantities can be handled at any one time. Nevertheless they are particularly useful in producing a number of radionuclides of biologically important elements such as oxygen, carbon and nitrogen. These have short half-lives so some hospitals, such as the Hammersmith in London, have their own small cyclotron dedicated to producing isotopes for medical use (Figure 10.5). The original cyclotron at Hammersmith in London was the first in the world to be built for this purpose. It closed down in 1985 after 30 years' service.

## Generator methods

The generator enables short half-life radionuclides to be supplied without the need for an on-site reactor or cyclotron. The general requirement is that the radioisotope of interest should be the short-lived daughter of a relatively long-lived radionuclide from which it can be separated. The separation or 'milking' of the generator 'cow' can be done either physically or chemically and either continuously or, more usually, as required.

## The $^{99}$Mo–$^{99}$Tc$^m$ generator

There are a number of generator systems of importance in nuclear medicine. The following description will concentrate on the $^{99}$Mo–$^{99}$Tc$^m$ generator as this produces the most commonly used radioisotope, technetium-99$^m$ (the m stands for metastable). $^{99}$Tc has a half-life of $2.1 \times 10^5$ years so is no use as a radiopharmaceutical (Table 10.4).

**TABLE 10.4** The $^{99}$Mo–$^{99}$Tc$^m$ generator decay

| Parent | $^{99}$Mo | decays with emission of ß⁻, with half-life of 67 hours |
| Daughter | $^{99}$Tc$^m$ | is produced which decays by emission of γ, half-life of 6 hours |
| Product | $^{99}$Tc | which has a half-life of $2.1 \times 10^5$ (years) |

The parent molybdenum is obtained from a nuclear reactor, either by an (n,γ) reaction or by the fission of uranium. The molybdenum is prepared as ammonium molybdenate and absorbed on alumina in a glass or plastic column. The $^{99}Tc^m$ is formed as the pertechnetate ion. When a saline (NaCl) solution is passed through the column the chloride ions exchange with the pertechnetate ions but not the molybdenate ions. This is called an ion exchange and it is a common separation technique. The saline now contains sodium pertechnetate.

The column is said to be **eluted** by the **eluent** (saline) to give the **eluate** (sodium pertechnetate in saline). It is sterile and compatible with the blood. After the elution the $^{99}Tc^m$ activity builds up and the column can be re-eluted. The volume of eluate is between 5 and 25 ml depending on the manufacturer.

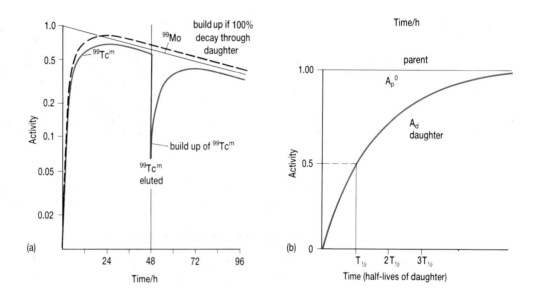

**FIG 10.6** Activity curves **(a)** $^{99}Mo$–$^{99}Tc^m$ **(b)** $^{113}Sn$–$^{113}In^m$.

**Yield** The activity of the daughter radionuclide at any time depends on:

**(a)** the activity of the parent

**(b)** the percentage conversion of the parent to daughter

**(c)** the time since the previous elution.

A typical activity of a $^{99}Mo$–$^{99}Tc^m$ generator will be calibrated as 5 GBq of $^{99}Mo$ at a particular time. This initial value $A_p^0$ will fall exponentially if the decay constant is $\lambda p$ so that the activity at time $t$ is:

$$A_p^t = A_p^0 e^{-\lambda_p t}$$

This represents the rate of formation of the daughter. If the decay constant for the daughter is $\lambda_d$, then the activity $A_d$, at time $t$ after an elution is given by:

$$A_d^0 (e^{-\lambda_p t} - e^{-\lambda_d t}) \tag{2}$$

This is shown for $^{99}Mo$–$^{99}Tc^m$ in Figure 10.6(a). In this system $\lambda_d > \lambda_p$ and not all the decay is via the metastable state, about 8.6 per cent decays direct to $^{99}Tc$. The dashed line is the curve of equation 2, as if all had decayed to $^{99}Tc^m$.

When $\lambda_d \gg \lambda_p$, the daughter has a very much shorter half-life than the parent, as in the $^{113}Sn$–$^{113}In^m$ system. The decay of the parent can be ignored in comparison

155

with the daughter, and equation 2 simplifies to:

$$A_d^t = A_p^0(1 - e^{-\lambda_d t})$$

which is shown in Figure 10.6(b). It can be seen that after several half-lives of the daughter, its activity approaches that of the parent. This build-up is exactly analogous to the production of a radionuclide by neutron bombardment in a reactor. In that case the neutron flux is constant, in this case the rate of decay of the parent is virtually constant.

### Continuous elution

In the $^{81}$Rb–$^{81}$Kr$^m$ generator, air passes over the column continuously removing the $^{81}$Kr$^m$. The activity of the daughter is straightforwardly calculated in this case; it is simply the activity of the parent multiplied by the decay constant of the daughter:

$$A_d = \lambda_d A_p$$

# THE CHOICE OF RADIOPHARMACEUTICALS

## 10.4

From the wide range of preparations available, the following factors will be considered in making the choice:

- The preparation must be **chemically available** in a high purity pharmaceutical form.
- The pharmaceutical must have a suitable **biological behaviour**, distributing itself in the organ or metabolic pathway in an efficient way.
- The radionuclide must have suitable **radiation characteristics** as regards radiation emitted and half-life.

## Chemical preparation

Radiopharmaceuticals must be prepared with the same attention to quality control as any other medicinal product, though their radioactive nature and, in many cases, short half-lives pose special problems.

There are two stages in their preparation: the preparation of the primary radionuclide, and the conversion of this into the chemical form required for the radiopharmaceutical use. At the end of the bombardment in a reactor it is usually necessary to separate the radionuclide from the target material and then carry out further purification. Methods of separation used have to be suitable for the small quantities usually involved, and the highly radioactive environment. Various chemical methods are used. Take the preparation of $^{131}$I as a typical example. The reaction is given in Section 10.3; at the end of the neutron irradiation, which lasts between one and four weeks, the target is removed and dissolved in 10 per cent sodium hydroxide. The solution is then acidified with sulphuric acid, ferric sulphate is added as oxidising agent, and the $^{131}$I is distilled off and collected in dilute sodium hydroxide. Alternatively the TeO$_2$ may be heated to drive off the $^{131}$I.

The formulation of a particular radiopharmaceutical can involve a wide variety of chemical reactions. Their availability is now greatly improved by the development of kits which contain all the ingredients necessary to prepare the radiopharmaceutical to the appropriate standard, without the requirement of extensive on-site facilities.

# Biological behaviour

The most important characteristic is that the radiopharmaceutical should localise and concentrate in the organ or system which it is required to image or treat. This maximises the effect required and minimises any side effects, which may be both physiological and radiological.

The normal metabolic activity of the body may result in a satisfactory concentration of the agent. The classic example of this is the concentration of iodine in the thyroid gland. This was in fact one of the earliest nuclear medicine procedures, and can be used both for imaging using scanners, and for therapy in the treatment of *carcinoma* (cancer) of the thyroid. The surprise is that although there are 22 isotopes of iodine to choose from, what is now usually used is $^{99}Tc^m$ as sodium pertechnetate. This is taken up in the same way as iodine but is more easily released.

Many other methods of localisation are used. Some are particularly effective in certain parts of the body. Diffusion is used in the brain; small particles in colloids can be ingested by the liver, others can be chosen so that they block the capillaries and lodge in the lung. One of the most recent developments results from genetic engineering techniques. These techniques make it possible to prepare substances called antibodies which target particular types of cell only. Tagging these with a radioisotope ensures that the radiation is delivered only to those cells. At high doses this can be used to kill tumour cells. Alternatively at lower doses, the radiation can be detected and used to measure the amount and distribution of a particular cell. This technique is called radioimmunoassay (RIA) and one of its uses is to measure growth hormone. Table 10.5 gives examples of commonly available radiopharmaceuticals, and their applications.

**TABLE 10.5** Examples of radionuclides used in imaging

| RADIONUCLIDE | RADIOPHARMACEUTICAL | ORGAN OR SYSTEM | FUNCTION |
|---|---|---|---|
| Technetium ($^{99}Tc^m$) | sodium pertechnetate | brain | brain blood flow |
| Technetium ($^{99}Tc^m$) | coagulated albumen particles | lungs | lung blood flow |
| Technetium ($^{99}Tc^m$) | colloidal suspension | liver | 1. liver blood flow<br>2. liver function |
| Technetium ($^{99}Tc^m$) | complex phosphate molecule | bone | 1. bone blood flow<br>2. bone metabolism |
| Technetium ($^{99}Tc^m$) | red blood cells | heart and circulation | 1. blood flow circulation<br>2. heart contraction |
| Iodone ($^{123}I$) | iodide | thyroid | thyroid metabolism |
| Iodine ($^{123}I$) | hippuran (sodium iodohippurate) | kidneys | 1. renal blood flow<br>2. renal function<br>3. urine flow |
| Xenon ($^{133}Xe$) | gas | lungs | lung ventilation |

# Radiation characteristics

**For imaging** X- or γ radiation of a suitable **energy** is required, particulate radiation is no use as it is absorbed within the patient. A single energy is preferred, not lower than about 100 keV, because of problems with increased scatter, and not higher than about 500 keV, as detectors become less efficient. The preferred type of radionuclide is one which decays by isomeric transition, such as $^{99}Tc^m$, as this maximises the amount of energy which is actually available for the imaging process. **Half-life** is an important factor; if the biological half-life is long then the physical half-life is the determining factor. This should be comparable to the time required to carry out the investigation to minimise the dose. When the biological half-life is of a similar period to that of the investigation, and the radionuclide is excreted, a long half-life radionuclide can be used.

**For therapy** beta emitters are often used, usually because they are absorbed in a small volume and therefore the dose is localised. The energy is not normally critical, though a reasonably high energy is required to give a significant penetration. A half-life of several days is preferred. Beta emission is often accompanied by gamma emission. The advantage of this is that the gamma radiation can be externally measured to monitor the dose. The disadvantage is the increased dose to the patient, and greater protection precautions required.

**TABLE 10.6** Properties of $^{99}Tc^m$

| | |
|---|---|
| Decay mechanism: | isomeric transition |
| Emission: | 140 keV gamma |
| Half-life: | 6 hours |
| Availability: | widespread from the $^{99}Mo$–$^{99}Tc^m$ generator |
| Chemical behaviour: | can be incorporated into a wide range of pharmaceuticals |
| Cost: | £10–£30 per dose |

### QUESTIONS

**10.11** Describe the production of a radiopharmaceutical of iodine-131, stating what precautions should be taken to ensure that the products are sterile. What 'special problems' would have to be overcome in preparing a product which is radioactive and which has a short half-life?

**10.12** Table 10.6 lists the properties of $^{99}Tc^m$, which is the radionuclide most commonly used in nuclear imaging. Explain why it is such an ideal radionuclide.

# NUCLEAR MEDICINE–RADIONUCLIDES IN ACTION

**10.5**

**T**he use of radionuclides falls into two major categories of therapy and diagnosis. The latter uses are often categorised as *in vivo* or in the body, which covers imaging and other tracer investigations, and *in vitro* or out of the body, which includes sample-taking investigations. Uptake tests are used to look at the functions of organs and body systems, dilution analysis enables measurement of composition of the body. In this section we will look at examples of each of these.

## Therapy

Radiotherapy is more commonly carried out using external radiation, usually from X-rays or cobalt-60 γ rays, as described in the previous chapter. The use of unsealed, liquid sources in the treatment of disease, is important in a few, specialised instances. Iodine-131 is taken orally in moderate doses (150–400 MBq), to treat overactive thyroids and in much larger doses (1–3 GBq), in the treatment of some cancers of the thyroid. Phosphorus-32 and yttrium-90 have been used to treat the blood and the pleura, respectively. The effectiveness of the treatment depends on targeting the affected organ or tissue with high doses of radiation; for this very significant localisation is required. This is the potential of the tagged antibodies mentioned earlier; they are the subject of much current research.

---

**QUESTION**

**10.13** Find out about the present state of the use of these antibodies in treating cancer by using a web browser.

---

## Thyroid function

This is an example of a dynamic function study, which is a common use for radioisotope tracers within the body, one which cannot be carried out by methods of external irradiation. Iodine-123 is given to the patient orally, usually as sodium iodide solution. It is absorbed from the gastrointestinal tract into the bloodstream and collected by the thyroid where it is used to produce hormones. Iodine-123 is a gamma emitter with energy of 160 keV and a half-life of thirteen hours. This is preferred to the beta emitting iodine-131 with a half-life of eight days, because it results in a much lower dose. In fact the most common radionuclide used in thyroid investigations now is the ever-useful $^{99}$Tc$^m$ because it stays in the thyroid for less time.

The emission can be imaged with a gamma camera or measured with a suitably collimated scintillation counter (see Chapter 12 for details). The count rate is then compared with a standard or model neck (sometimes called a 'phantom') containing the same amount of iodine-131. The uptake ratio

$$\frac{\text{count rate from patient's thyroid}}{\text{count rate from model or standard}}$$

is recorded at various intervals, usually over a 24-hour period. From the results it can readily be seen if the thyroid is over (hyper-)active, normal or under (hypo-)active.

## Lung imaging

Two types of lung function studies are used, one of blood flow called **perfusion**, the other of air flow (**ventilation**). For perfusion, albumen, which is part of human blood plasma, can be coagulated (like egg-white during cooking), suspended in saline and labelled with $^{99}$Tc$^m$. This is then injected into the bloodstream where it is trapped in the fine capillaries of the lung. A map is produced of the functioning capillaries and so represents the distribution of blood flow (perfusion) to the lungs. If an *embolus* (blood clot), formed elsewhere in the system, reaches the lungs it can block the blood supply with potentially fatal consequences. This area will show on the image as a region without tracer. In a ventilation study, the patient breathes in oxygen which is mixed with the radioactive gas $^{133}$Xe (xenon), and a gamma camera produces an image of the activity which results from the presence of the radionuclide in the airspaces of the lung.

## Bone function

The turnover of phosphate by the bones can be imaged by $^{99}Tc^m$ labelled to a complex phosphate molecule. If a cancer has produced secondary deposits in the skeleton, these show up as increased uptake, or hot spots. Such information about function can be detected much earlier than with a radiograph which is better for structure of the skeleton, and this can be important in treating the patient (Figure 10.7).

## Measurement of body fluids

Following an accident in which there has been significant, but unknown loss of blood, it is desirable to measure what remains in the patient. Since it is not possible to simply drain this out and measure it (!), the technique of **dilution analysis** is used. About 200 kBq $^{131}$I-labelled albumen in saline is injected into one arm. After about 15 minutes a blood sample is drawn from the other arm. Since the count rate of the injected material is usually too high to measure directly on the same instrument for direct comparison, it is usual to dilute an equal amount of the radioactive material with a known volume of water, and count a sample after mixing thoroughly.

> **Example**: A patient is injected with 5 cm$^3$ of $^{131}$I-labelled albumen. 15 minutes later a 5 cm$^3$ sample is drawn from the patient and found to have an activity of 100 Bq. Another 5 cm$^3$ sample of the radiopharmaceutical is mixed with 1000 cm$^3$ water and the activity of this is measured to be 500 Bq. What is the patient's blood volume?
>
> Original activity of sample $= \dfrac{1000}{5} \times 500\ \text{Bq} = 10^5\ \text{Bq}$
>
> The patient's blood has diluted the activity by $10^5/100 = 1000$, so the volume of his blood is 1000 times greater than the injected volume, i.e. 5000 cm$^3$.

A similar technique is used to measure total body water using tritiated water, that is, water containing the radioisotope tritium $^3$H.

Other examples of *in vivo* investigations are summarised in Table 10.5, along with those described here. There are constant developments to enable new processes and structures to be imaged, to improve those already imaged and to reduce the dose. One way to do this is the *in vitro* test, so called because it is carried out in a test tube. The radioisotope is added to the sample, of blood for example, after it has been taken from the body.

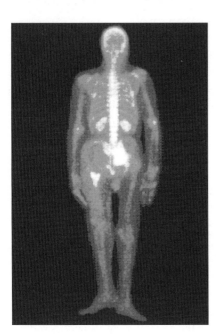

**FIG 10.7** A gamma camera scan of a person with secondary cancer in the spine, shown by light spots where the radioisotope has been preferentially absorbed by the cancer cells.

## SUMMARY ASSIGNMENTS

**10.14** **(a)** Define the terms *physical half-life* and *biological half-life*.

**(b)** Chromium-51 is used to measure the red blood cell volume in a patient. The physical half-life of chromium-51 is 28 days. For a normal patient the biological half-life of the cells labelled with this isotope is 35 days.

    **(i)** Calculate the expected effective half-life of this isotope when given to a normal patient.

    **(ii)** If the measured effective half-life in a patient with a blood disorder is 4 days, calculate the biological half-life of the labelled cells for this patient.

*(AQA June 1999)*

**10.15** Iridium-192, in the form of a thin wire, is placed inside a patient's body during the treatment of certain medical conditions. Both β radiation and γ radiation are emitted by the iridium.

(a) Iridium-192 decays to an isotope of platinum by beta decay. Write down in the nucleon and proton numbers for the isotope of platinum in the equation below.

$$^{192}_{77}\text{Ir} \rightarrow \text{Pt} + \beta^- + \gamma$$

(b) What precautions against the dangers of ionising radiation would the medical team who carried out the treatment need to observe?

(c) The half-life of beta decay for iridium-192 is 75 days. Calculate its decay constant.

(d) Show that the number of atoms in a 10 mg sample of iridium-192 is $3.1 \times 10^{19}$.

the Avogadro constant $= 6.02 \times 10^{23}$ mole$^{-1}$

(e) Calculate the activity of this sample immediately after it is prepared.

(f) How many atoms of iridium-192 would there be left after 30 days?

**10.16** This question is about the use of radioactive tracers in the human body.

(a) What is a radioisotope? Discuss the factors which affect the choice of a radioisotope for a particular medical investigation.

(b) Describe how a radioisotope *of your choice* is used in a particular investigation. You should name the isotope used, state what radiation is emitted and how this is monitored. You should also describe what measurements are made and how they are used to derive a result, mentioning any precautions or relevant safety procedures.

*(O & C June 1997)*

**10.17** (a) State what is meant by the terms *radioactive half-life* and *biological half-life*.
Why does biological half-life depend on both the organ and the patient?

(b) Outline the use of a radioactive tracer in measuring the volume of blood in a person. You may be awarded a mark for the clarity of your answer.
Iodine-131 has a half-life of 8 days. Approximately what percentage of a sample of iodine will remain in the body after a period of one month?

Complete the following equation for the decay of iodine-131.

$$^{131}\text{I} \rightarrow {}_{54}\text{Xe} + \beta^- + \gamma$$

A second isotope of iodine, iodine-123, is available. It decays by emission of gamma radiation only.

Explain which isotope of iodine you would prefer for treatment of an over-active thyroid gland.

*(Edexcel Specimen 2001/2)*

# 11

# MEASURING AND CONTROLLING RADIATION

**T**he mushroom cloud of a nuclear explosion has come to symbolise the dilemma of our use of nuclear changes. The nuclear 'genie' offers the power to improve or destroy our environment, to kill or cure the human race. The difference between the two extremes may only be a matter of quantity. How much energy and radiation is released, over what period and in what way? These are important questions which this chapter is designed to help answer. You will study how the important aspects of radiation are quantified, measured and controlled.

## LEARNING OBJECTIVES

*After studying this chapter you should be able to:*

① define, know the units of, and use in calculations, the following terms: activity, exposure, absorbed dose, equivalent dose, exposure rate constant;

② use the following terms correctly: committed equivalent dose, effective dose, collective dose, linear energy transfer, quality factor, relative biological effectiveness;

③ recall the background levels of radiation, and give examples of some common medical doses, such as a chest X-ray;

④ discuss, in general terms, and by giving examples, the relationships between radiation levels and the incidence of damage or disease;

⑤ recall the general principles under which the use of radioisotopes is permitted, and give examples of maximum permitted dose levels.

# BIOLOGICAL EFFECTS OF RADIATIONS

**A**ll types of radiation carry energy. Their biological effects are caused by the way that they deposit that energy in tissue. Radiations in the lower frequency regions of the electromagnetic spectrum dissipate their energy at the surface of the body: in strong sunlight we experience a heating and a reddening of the skin. The radiations we are concerned with in this theme can penetrate the body tissues to varying degrees and cause ionisation. The general way in which this leads to damage was described in Section 9.2 and a summary of the effects at different levels of biological organisation was given in Table 9.1. As a preliminary to quantifying these effects, let us consider what factors will affect the damage which these radiations cause.

## Type of radiation

The interaction of $\alpha$, $\beta$ and $\gamma$ with body tissue was described in Section 10.2 and schematically in Figure 10.4. The significant difference is in the extent of the ionisation produced, which affects the penetration into the tissue. The penetration of alpha particles into the skin is only about 50 $\mu$m whereas a high proportion of gamma radiation will pass completely through the whole body.

---

### QUESTION

**11.1** (a) Explain why sealed $\alpha$ sources, such as those used in physics laboratories, are unlikely to cause damage, while an open $\alpha$ source, such as a liquid or gas is potentially very hazardous.

(b) Why is the interaction of $\gamma$ radiation with tissue very similar to that of X-rays?

(c) In what ways is the interaction of $\beta$ radiation with tissue intermediate in its effect, between $\alpha$ and $\gamma$? Explain why this is so, with reference to Fig 10.4.

---

## Activity of the source

As the activity is the measure of the number of disintegrations per second, this is clearly important in determining the effect of a particular source. Radiopharmaceuticals will always have a quoted or reference activity, as do laboratory sources. This represents the maximum activity; decay will result in a falling activity with time according to the equation

$$A = A_0 e^{-\lambda t}$$

It may be very difficult to determine the activity in a particular situation since all the relevant information may not be readily available. Take the example of radioactive fallout from the Chernobyl nuclear reactor, which reached Britain on May 2–4 1986. Estimates made of the activity of the radioactive contamination of Glasgow a few days after the Chernobyl accident, varied by a factor of four times, largely because of the assumptions made about rainfall (Figure 11.1).

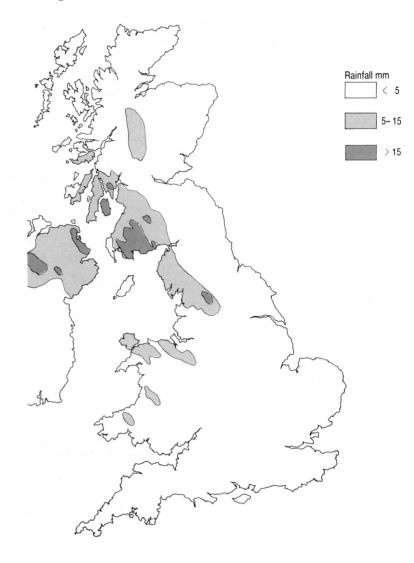

Rainfall mm

☐ < 5

▨ 5–15

▩ > 15

**FIG 11.1** Rainfall in the UK during the passage of the Chernobyl plume 2–4 May 1986. This is not the total rainfall for that period, only that which intercepted the plume.

## QUESTION

**11.2** Briefly explain how the areas of Britain most polluted by the radioactive fallout from Chernobyl were those where it rained whilst the radioactive cloud was passing over. How would you expect this activity to cause exposure to humans? What would be the best way to measure it? How do you think a difference of a factor of four might arise in the estimates?

## Time of exposure

For a given source of a particular activity, the exposure to radiation will be directly proportional to the period of time. With sources of short half-life, clearly the activity does not remain constant over long periods, and this will have to be taken into account. The biological effect of the exposure does depend on the period over which it is received. A number of small doses, or a continuous low level of irradiation is less damaging than the same dose given over a short period of time. This is because cells are able to recover from the damage caused when this is small.

As a consequence of this, there are different dose limits for short and long periods of exposure.

## Type of tissue irradiated

It was discovered as long ago as 1906 that cells differed considerably in their sensitivity to radiation. As a result of their studies, Bergonie and Tribodeau concluded that rapidly dividing or mitotically active cells are radiosensitive; mature, differentiated cells which are unlikely to undergo cell division are not. This means that those tissues which most rapidly replace themselves are the most susceptible to damage; the skin, bone marrow and reproductive organs. It also accounts for the ability of radiation to cure cancers as well as cause them. What distinguishes cancerous cells from normal ones is that they divide more rapidly. The least affected cells are those of the brain and bones.

## 11.2 DOSIMETRY

**T**he quantification of radiation effects is complicated for three reasons:

1. The effects of the radiation's interactions with tissue is complex and not fully understood.
2. There are many relevant quantities which we need to measure.
3. There are at least two systems of units; in practical dosimetry the SI system of units has not yet completely replaced earlier systems.

In this section we will consider the quantities to be measured and the various units in common use. Some of the complexities and uncertainties will be considered in the next section.

## Activity

This is simply a measure of the number of nuclei that disintegrate in one second. The unit the becquerel, Bq, is one disintegration per second. This is a very low activity, typical values of sources used in the teaching of physics are about one MBq, medical sources are often in the GBq ($10^9$ Bq) range. The old unit of activity is the curie, Ci which is equivalent to $3.70 \times 10^{10}$ Bq

The curious (!) conversion from curie to becquerel arises because the curie is defined as the activity of one gram of radium. What does this tell you about the activity of radium? How did the curie get its name?

## Absorbed dose

The origin of radiation damage is in the energy given to the irradiated material, so the absorbed dose is simply the energy absorbed per unit mass:

$$D = \frac{E}{m}$$

The unit is therefore J kg$^{-1}$ which is called the gray, Gy. The old unit, the rad, derived from the metric 'cgs' system; 1 rad = 0.01 gray.

## Exposure

With X- and γ radiation it is found that the total ionisation produced is a good measure of the total energy absorbed. Exposure is defined as the total charge of one sign (+ or –) produced in a unit mass of air:

$$X = \frac{Q}{m}$$

The unit is therefore C kg⁻¹. The old 'cgs' unit of exposure is the roentgen, R; $1R = 0.01$ C kg⁻¹.

Charge is usually easier to measure than energy, so this is a useful quantity. It applies, strictly speaking, only to X- and γ radiation in air. The relationship between exposure and absorbed dose in other circumstances can however be compared to this.

## Relationship between exposure and absorbed dose

An exposure in air of 1 C kg⁻¹ means the creation of $1/e$ electrons per kilogram of air, where $e$ is the electronic charge. Each of these requires an energy, the ionisation energy, of 34 eV, so the total energy released is:

$$34e/e = 34 \text{ J kg}^{-1} = 34 \text{ Gy}$$

so for air:

$$D \text{ [Gy]} = 34 \times X \text{ [C kg}^{-1}]$$

and in general:

$$D \text{ [Gy]} = f \times X \text{ [C kg}^{-1}]$$

where $f$ is a conversion factor which depends on the absorbing material. The atomic number Z of soft tissue is similar to that of air (Z for air is 7.6, Z for muscle is 7.4), so $f \approx 34$ J C⁻¹. When the photon energy of the radiation is high, energy absorption is by Compton scatter which is independent of Z, so the relationship between exposure and absorbed dose is constant for all materials. Otherwise $f$ varies with photon energy and the relationship can be computed with the use of a graph such as Figure 11.2.

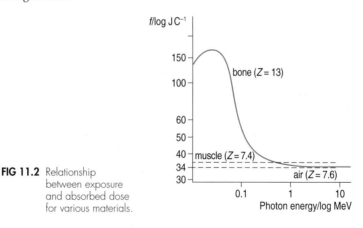

**FIG 11.2** Relationship between exposure and absorbed dose for various materials.

## Equivalent dose

Radiation damage depends on the distribution of the energy released, and this in turn depends on the type of radiation. Those with multiple charge and multiple atomic mass units, such as α, deposit their energy in a short path length. They are called high linear energy transfer radiations. This effect is expressed as a dimensionless constant, the quality factor, Q, or relative biological effectiveness of the radiation. The values recommended by the International Commission on

Radiological Protection, the ICRP are given in Table 11.1. The equivalent dose of a particular radiation is defined as:

$$H = QD$$

The basic unit is J kg$^{-1}$ but it is called the sievert, Sv, to distinguish it from the absorbed dose. The old unit is the rem; 1 rem ≈ 0.01 Sv. Figure 11.3 shows the relationship between these quantities, in a schematic form.

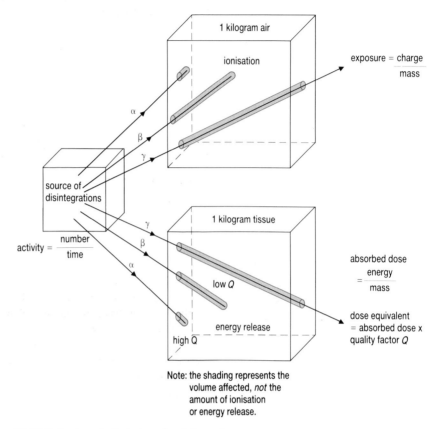

**FIG 11.3** A schematic illustration of radiation units.

The linear energy transfer LET, of a charged particle is the average energy transferred to the absorbing medium, per unit distance travelled.

**TABLE 11.1** Quality factors

| RADIATION | Q |
| --- | --- |
| X-rays, γ rays, high energy ß | 1 |
| 'Thermal' neutrons | 5 |
| Other neutrons and singly-charged particles | 10 |
| α particles and multiply-charged particles | 20 |

### Some UK values

Annual doses from radiation of natural origin. Figures are averages for the whole population and are approximate (from NRPB 1987).

| | |
| --- | --- |
| Effective dose | 1870 mSv |
| Genetically significant dose | 1000 mSv |
| Collective effective dose | 103 kman Sv |

A number of other measures are in common use. The **committed equivalent dose**, is the total equivalent dose over the period of time that a radionuclide is in the body. It will depend on the physical and the biological half-lives. The **effective dose** is a measure which takes account of the relative sensitivities of different parts of the body. Each is assigned a particular weighting so that a total can be made from whatever parts of the body are irradiated, which represents the total effect of the radiation, as if the whole body was irradiated uniformly. Of particular concern is that part of the radiation which may affect the reproductive organs, or gonads. This is called the **genetically significant dose**.

It is often necessary to calculate the total equivalent dose of a group of people, for example, those who live near a nuclear power station. The **collective equivalent dose** is simply the average equivalent dose, multiplied by the number of people exposed. It is usually expressed in man sievert although those committed to anti-sexist language refer to the person sievert.

## Exposure rate constant

The equivalent dose can be calculated directly from the activity of a source, by the use of this constant, R which represents the average energy release for the particular radiation of a particular nuclide. This is used for example in estimating the maximum dose during the use of sealed sources in physics laboratories. The constant is expressed in terms of the dose received, in mGy, at 1 metre over a period of 1 hour, from a source of 1 MBq activity. The constant therefore has a rather cumbersome unit, for example, $\gamma$ radiation from Co-60:

$$R = 0.36 \text{ mGy m}^2 \text{ MBq}^{-1} \text{ h}^{-1}$$

## Calculations

Here is a worked example using the rate constant.

A student is carrying out an experiment with a Co-60 source of labelled activity 5μCi(~0.2 MBq). She holds the source in tongs 25 cm long for the duration of the experiment which is 6 minutes. What is the maximum equivalent dose that she could receive?

$$D = \frac{R \times A \times t \text{ [hours]}}{d \text{ [metres]}^2}$$

$$D = 0.36 \times 0.2 \times 0.1/(0.25)^2 \text{ mGy}$$

$$= 0.072/0.625 = 0.115 \text{ mGy}$$

This is the maximum dose because (i) there will be some shielding from the source container, (ii) only the nearest part of the body will have been exposed to this degree, (iii) the source will have decayed to below its maximum activity, $T_{1/2}$ for Co-60 = 5.3 years.

The equivalent dose is 0.115 mSv because the relative biological effectiveness of $\gamma$ radiation is 1. We will see in the next section how this compares with a 'safe' dose.

---

### QUESTIONS

**11.3** (a) Suppose the student in the previous example held the source in her hand instead of tongs (this is against the regulations!), so that the distance from source to hand is reduced to 0.5 cm.
What is her equivalent dose now, assuming the experiment still lasts 6 minutes?

(b) Suppose that a student was foolish enough to put such a source in his pocket for an hour; what dose would he receive if the distance between source and body is 1 cm?

**11.4** How much energy is absorbed when a student of mass 60 kg receives an effective equivalent dose of 20 mSv, half from radiation of quality factor 1 and half of quality factor 2?

# RADIATION LEVELS

## Incidence of radiation

The responsibility for monitoring the extent of radiation exposure and setting recommended dose limits is with the ICRP, internationally, in conjunction with national organisations such as the National Radiological Protection Board (NRPB) in Britain. Their recommendations usually become the safety regulations, and these are discussed in the next sub-section. Both the monitoring and regulatory functions lead to changes in the estimated doses; see for example the case study box and question 11.8 on page 173. Other organisations may contest the risk estimates of the 'official' figures. Whenever you meet data on this subject, it is worth noting who published it, if it is their own work and when the work was carried out. This may help you to decide why data does not agree.

## Natural background radiation

This originates from the action of stars which are vast nuclear reactors. Cosmic radiation is the direct result of this, mainly of course, from our own Sun. The Earth itself originates from this, so there are many naturally radioactive substances in the Earth, such as compounds of uranium, potassium and thorium. Some of these decay to form different radioisotopes which, in the case of radon and thoron pollute the atmosphere, particularly in buildings made from radioactive minerals. Concern about this has grown in recent years; NRPB estimates that it is the second most important cause of lung cancer in the UK, after cigarette smoking. Accurate estimates are difficult to make as the dose is very dependent on air flow, so monitoring is continuing. Potassium-40 is a common radionuclide in the body and the main source of internal radiation.

**TABLE 11.2** Annual dose of radiation of natural origin in mSv *Nature* 16.3.1989

| SOURCE | UK (1988) | US (1987) |
|---|---|---|
| Radon and thoron from earth | 1.30 | 2.00 |
| Terrestrial gamma ray | 0.35 | 0.28 |
| Internal radiation | 0.30 | 0.39 |
| Cosmic radiation | 0.25 | 0.27 |
| Totals | 2.20 | 2.94 |

### Risks from radon

A lifetime exposure to radon at the 'action level' of 200 Bq m$^{-1}$ will give rise to a 3–5% risk of an individual dying of lung cancer

*Health risks from Radon NRPB, 2000.*

The values quoted in Table 11.2 are averages for the country's population. The values are higher in regions of granite rocks (more uranium), in regions of higher altitude and latitude (less protection from cosmic radiation by the atmosphere), and in draught-proofed, centrally-heated homes. The variations in background radiation are, as you would expect, much greater worldwide. The maximum that is known for a large population is the 100 000 or so inhabitants in part of the Indian states of Kerala and Madras. As a result of the underlying rock they receive an average dose of 5–10 mSv per year; about four to eight times the world average dose. No significant increase in radiation-linked disease has been unequivocally associated with this population or any other, as a result of variations in background incidence (though it might be discovered yet!). Some worldwide figures are given in Table 11.3.

**TABLE 11.3** Worldwide examples of annual doses from natural sources (various sources)

| REGION/CITY | POPULATION | DOSE/mSv |
|---|---|---|
| London, UK | $11 \times 10^6$ | ~1 |
| New York, USA | $17 \times 10^6$ | ~1 |
| Paris, France | $10 \times 10^6$ | ~1 |
| Central France | $7 \times 10^6$ | ~3 |
| Parts of Cornwall, Yorkshire and Aberdeen, UK | $4 \times 10^5$ | ~5 |
| Kerala and Madras, India | $1 \times 10^5$ | 5 to 10 |

## Artificial radiation

The largest source by far is medical uses as Table 11.4 shows. These values are averages for the whole population. In the case of medical uses this includes diagnostic X-ray, in which a large number of people will have small doses (0.1–2 mSv), and radiotherapy, in which a small number will have very large doses (30–70 Sv). In the 1990s the dose due to X-ray diagnosis has been reduced quite significantly by newer, more sensitive techniques requiring about 30% less exposure and by its replacement with ultrasound examinations. However there has been an overall increase in the average dose for medical use due to the increase in CT scanning procedures. NRPB estimate these now make up about 40% of the collective dose, though they account for only 4% of the procedures. Some specific medical doses are listed in Table 11.5.

**TABLE 11.4** Annual dose of radiation of artificial origin in µSv
*Nature* 16.3.1989

| SOURCE | UK (1988) | US (1987) |
|---|---|---|
| Medical uses | 300 | 530 |
| Fallout (including Chernobyl) | 10 | 0.6 |
| Miscellaneous sources | 10 | ~100 |
| Occupational exposures | 5 | 9 |
| Radioactive waste | 1 | 0.5 |
| Totals | 326 | ~640 |

**TABLE 11.5** Example of typical medical doses in 1990s (NRPB)

| SOURCE | DOSE/µSV |
|---|---|
| Dental X-ray | <10 |
| Chest X-ray | 20 |
| Pelvic X-ray | 70 |
| 'Barium meal' | 3000 |
| CT scan head | 2000 |
| CT scan chest | 8000 |
| Thyroid scan $^{99}Tc^m$ | 1000 |
| Heart function $^{99}Tc^m$ | 1000 |
| Lung perfusion $^{99}Tc^m$ | 1000 |
| Lung ventilation $^{81}Kr^m$ | 100 |

Other sources make a small contribution on average, but this may in reality be a large dose for a small number of people. Occupational doses are those received by people who work with radioisotopes, again a small proportion of the total population. Miscellaneous sources include television sets, fire alarms which contain α sources and work like a cloud chamber, gas mantles containing thorium and liquid crystal watches which contain tritium. These latter have replaced luminous watches which had radium in the painted numbers. Those who painted the watches made a fine point on their brushes by licking. After some years it was realised that this caused high levels of radiation-induced diseases, from the ingestion of radium.

## INVESTIGATION

### Background radiation

Make measurements of the normal background radiation in your locality, using a Geiger-Muller tube and a scalar/counter. This will give you a result in counts per unit time. Now compare this with the radiation from other sources available. Possibilities are small quantities of uranium and thorium salts, a gas mantle (these are sold in camping shops), and about 500 grams of a potassium salt (this is about the quantity of potassium that you have in your body).

Try to obtain the best estimate of the background level of radiation for your area from reference materials available to you. You can then calibrate your first measurement so that you can obtain values of the other source doses in grays.

## QUESTIONS

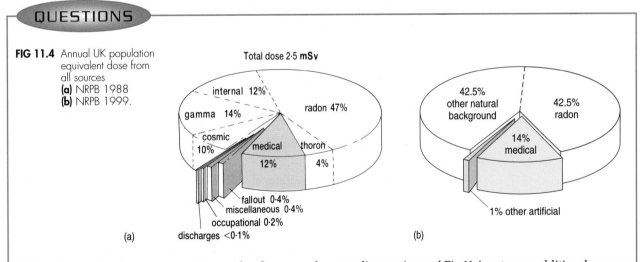

**FIG 11.4** Annual UK population equivalent dose from all sources **(a)** NRPB 1988 **(b)** NRPB 1999.

Total dose 2·5 **mSv**

(a) internal 12% · gamma 14% · cosmic 10% · medical 12% · thoron 4% · radon 47% · fallout 0·4% · miscellaneous 0·4% · occupational 0·2% · discharges <0·1%

(b) 42.5% other natural background · 42.5% radon · 14% medical · 1% other artificial

**11.5** Answer the following questions with reference to the preceding section and Fig 11.4, or to any additional sources of information. For example the NRPB website or internet sites for environmental pressure groups.
   **(a)** What changes in medical procedures occurred in the period 1988 to 1999 which would account for the differences between Fig 11.4 (a) and (b)?
   **(b)** What can be done to reduce exposure to other artificial sources of radiation? What estimates of risk reduction follow from this? How much do views vary in this?
   **(c)** Find out why the miscellaneous sources mentioned are radioactive and list any others you know of.
   **(d)** Explain how the natural background radiation varies within a single country. Why would you expect the worldwide variation to be much greater?

**11.6** Is your own dose of radiation in the past year likely to be more or less than the average given? Think about the various sources and your own experiences, such as any medical X-rays, holidays in areas of high natural radioactivity, and so on. Can you make a quantitative estimate?

## What level of radiation is safe?

The answer you gave to the last question was probably no, but if you could estimate the dose, how would you know if it was safe? This is not an easy question as you no doubt realise! The fundamental problem is that so many parts of the damage mechanisms described in Section 9.2 seem to depend on probabilities, not certainties. Like the process of radioactive decay, the results can only be described with any accuracy if there are a large number of instances. This is not yet the case for damage at high or moderate levels of radiation. A further complication in estimating the effects is that these do not cause disabilities or deaths which are uniquely identifiable with radiation exposure. Cancers for example are caused by smoking, dietary actions, atmospheric pollutants and other environmental causes, to which low-level radiation only adds a small proportion. Most of the existing data has come from the survivors of the Second World War bombing of Hiroshima and Nagasaki. Studies of the effects of the Chernobyl reactor fallout in 1986 are still continuing. At the time 237 cases of acute radiation sickness, including 31 deaths, were reported.

## 11.4 RADIATION PROTECTION

The system of protection, which is to a large extent internationally agreed involves a set of principles leading to regulations covering the use of radioactive substances in all circumstances.

## Principles

The radiation protection agencies set maximum permitted dose levels (MPLs) for various situations. The principles on which these are based are:

- **Limitation** There is no threshold, or safe dose – a linear relationship is usually assumed for stochastic effects.

- **Optimisation** All additional doses should be as low as is reasonably achievable, (called ALARA).

- **Justification** Any use should show a positive net benefit.

Each of these is difficult to quantify and involves some value judgements. Thus it is to be expected that people will disagree about the extent to which radiations should be used in any particular circumstance. One of the more straightforward choices is in the use of radiotherapy in treating a fatal disease. Perhaps only one individual's risk and benefit has to be considered. The case reported in the margin indicates how agonisingly difficult even this can be.

One of the consequences of this approach is the setting of the *different* dose limits. For a radiation worker this is 20 mSv per year for the reproductive organs or the whole body, rising to 500 mSv per year for the limb extremities. For the lay person it is 1 mSv per year and for children under the age of 16, 0.5 mSv per year. As more data become available and as understanding of the damage mechanisms increases it is to be expected that dose limits will change.

A man in his late sixties was in terminal care from a tumour called pineoblastoma. His conventional radiotherapy treatment had blinded him; he couldn't move, he had lost weight. He was dying. Three weeks after a new treatment involving a targeted antibody carrying radioactive iodine he was able to walk out of hospital, free of pain but still blind. He survived two years with no problems. He relapsed and though he could have had another injection he was so depressed by his blindness that he refused further treatment.

*The Guardian* 20.4.1988

QUESTION

**11.7** (a) What are the risks and benefits of a dental X-ray procedure? How can the dose be made ALARA?

(b) What is the justification for setting a higher limit for radiation workers?

(c) Why are these workers permitted a higher dose to their hands and feet?

# Regulations

Any medical use of radioactive substances or equipment producing radiations, must comply with the relevant regulations. These are usually set by national government departments on the advice of bodies such as ICRP and, in Britain, NRPB. Radionuclides are divided into classes by their toxicity (ability to harm). In addition to setting limits on the amounts, the regulations for each class deal with three aspects of safety: people (what are the necessary qualifications, duties and responsibilities), places (for which buildings, services and purposes), and procedures (how to do things safely).

**Radiotoxicity** The first distinction to be made is between sealed and open sources, as the hazard of contamination is absent from sealed sources. The sealed sources have the highest activity, cobalt-60 sources for teletherapy can be as high as 300 TBq ($3 \times 10^{14}$ Bq). The highest levels of open sources encountered in hospitals is iodine -131 for the therapy of thyroid cancers, having activity of about 5 GBq. Some hospitals require nuclides with activities of only tens of MBq. The nature of the radiopharmaceutical is important in determining the toxicity, as well as the activity, since this determines how it affects the body chemically.

## CASE STUDY

### Radiation more hazardous than previously thought

RADIATION is two or three times more dangerous than was previously thought, the National Radiological Protection Board said yesterday. The board, a Government watchdog, wants the limits for maximum legal radiation doses to be lowered accordingly.

The recommendations could affect about 2,000 people whose work takes them near the present dose limits. It will increase pressure on British Nuclear Fuels and the UK Atomic Energy Authority to lower radioactive discharges from the nuclear reprocessing plants at Sellafield and Dounreay.

Dr Roger Clarke, the director of the Protection Board, said a radiation worker exposed to the maximum legal dose of 50 millisieverts (mSv) a year had been thought to face a one in 2,000 risk of contracting fatal cancer each year. Now the risk is put at one in 700.

'Continued exposure near the (current) dose limits represents a level of risk which verges on the unacceptable,' Dr Clarke added.

Such exposure equates to roughly a one in 18 chance of contracting a fatal cancer during a 40-year working life.

Dr Clarke said people working near the dose limits included some maintenance workers at Sellafield plant, Ministry of Defence staff working on nuclear weapons, industrial radiographers checking pipe welds and surgeons and theatre nurses exposed to radiation beams during operations. 'They form about one per cent of those workers whose exposure to radiation is monitored – about 2,000 in the whole country,' Dr Clarke added.

The NRPB suggest that employers should lower the maximum dose for workers to an average of 15 mSv a year while the Government's Health and Safety Commission considers reducing the legal limit.

The board's stance follows a decision by the International Commission on Radiological Protection to accept the findings of a definitive study, on survivors of the Hiroshima and Nagasaki atom bombs. The indicated radiation was more likely to cause fatal cancers than thought previously. While the Commission decided a change in the dose limits was not necessary, the NRPB disagrees and is breaking the international consensus.

The board recommends that no member of the general public should be exposed to an artificial radiation dose of more than 0.5 mSv a year. The present legal limit is 1 mSv. Under the new risk estimates, this equates to a one in 33,000 chance of contracting a fatal cancer each year.

*The Independent 19.11.1987*

---

### QUESTION

**11.8**  Read the passage in the box and answer these questions on it.
  **(a)** Who are the people whose jobs put them at a risk of a one in 700 chance of contracting a fatal cancer each year?
  **(b)** Do the reductions in the MPLs recommended in the article correspond with the opening statement?
  **(c)** Try to find out the total risk in the general public of contracting a fatal cancer each year, hence work out what proportion of this would be added by a dose of 1 mSv.

---

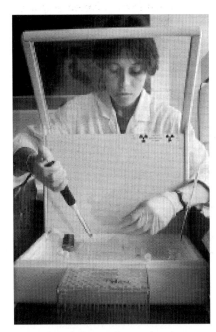

**FIG 11.5** Safety procedures with radioactive tracers, here used for labelling in immunological research.

**People**  In general, regulations will say who is responsible for what, when X-rays and radionuclides are being used. In Britain, the Health and Safety at Work Act gives some responsibility to the individual worker as well as the employer. The regulations for work with each class of radionuclide identify a Radiation Protection Advisor (RPA), whose responsibility it is to draw up local rules. These are implemented at a given workplace, by the Radiation Protection Supervisor (RPS). Each person who initiates an activity involving the use of radiation, for example, the radiologist, is also required to take additional responsibility. So the chain of responsibility goes from the general to the particular.

**Places**  Approval for the use of X-rays and radionuclides is given to particular places, subject to the appropriate people being available to carry out the required procedures with suitable facilities. For example, a hospital with a reasonably large diagnostic facility would have a specialist medical physicist as RPA with specialist radiographers as RPSs. The required facilities might be those listed in Table 11.6. Areas containing any sources have to be labelled, and some areas have limited access to staff and patients. Monitoring of exposure levels in these sites, and of the personnel entering them is required (Figure 11.5).

**Procedures**  These are the detailed accounts of what is expected of those people with the various levels of responsibility described above. Those with the more widespread responsibilities will have the most detailed procedures to follow. The general purposes are to establish and maintain practices which keep radiation exposure as low as possible, and within any limits set. For the individual employee, this will mean following rules similar to those you have when using sources in physics. These may be summed up as:

> When using radioactive sources, increase your **distance** to a maximum, reduce your **time** to a minimum, use **shielding** where appropriate, and with open sources, avoid or contain any **contamination**.

A supervisor's procedures will include instructing and monitoring all staff, monitoring the performance of equipment, keeping records of staff, patients and equipment, and perhaps assessing new equipment and evaluating the changes needed in the safety procedures as a result of this.

**TABLE 11.6** Facilities of a laboratory for handling radionuclides

Specifically set aside for handling radionuclides.

Protective clothing used. Gowns or overalls, gloves and overshoes.

Negative pressure. Extraction fans to outside of building to prevent spread of contamination inside building.

Fume cupboards.

Floors, walls and benches should be smooth and without cracks. Must have non-absorbent and washable surfaces.

Benches strong enough to support lead shielding.

Wash basin with elbow and/or foot operated taps.

Sink with direct access to main drain for disposal of radioactive liquid waste.

## SUMMARY ASSIGNMENTS

**11.9** **(a)** When calculating the radiation *dose equivalent* received by a patient a *quality factor* is taken into account. When considering patient safety explain the significance of *dose equivalent* and *quality factor*.

**(b)** A patient receives an absorbed dose of 40 mGy. Half of the absorbed dose comes from radiation of quality factor 1 and the other half from radiation of quality factor 20. The patient's whole body is exposed to the same level of radiation.

**(i)** Calculate the dose equivalent absorbed by the patient.

**(ii)** Calculate the total energy absorbed by the patient who has a mass of 60 kg.

*(AQA Mar 1999)*

**11.10** Cobalt-60 sources are used in medicine for *radiation therapy*.

**(a)** **(i)** State what is meant by *radiation therapy* and suggest a patient condition for which it might be used.

**(ii)** What type of electromagnetic radiation is emitted from cobalt-60?

**(iii)** Give **two** properties of cobalt-60 which make it a suitable source for use in radiation therapy.

**(b)** A cobalt-60 source gives a dose rate of 80 μSv per hour over an unshielded area of body when positioned 1.2 m from a patient.

**(i)** At what distance from the source would the patient receive a dose rate of 20 μSv per hour over the same unshielded body area?

**(ii)** The half-value thickness of lead used to absorb the radiation from cobalt-60 is 12.5 mm. If the source and patient are kept 1.2 m apart, what thickness of lead, placed between the source and the patient, would reduce the dose rate of 20 μSv per hour?

*(AQA Mar 1999)*

# 12 DETECTORS AND IMAGERS

The measurement and imaging of radiations is required for a range of purposes and in a variety of situations, as the previous chapters have shown. This has led to the development of many devices and techniques. In this chapter we will study the principles of these and look at examples of the commonest types in use in hospitals. The photographic, ionising and exciting effects of radiation form the basis of all methods of detection and imaging. The chapter describes each of these in turn and this is followed by an account of some of the current developments in imaging techniques. Table 12.1 summarises the methods and devices.

**TABLE 12.1** Methods and devices for detecting radiation

| METHODS | PHOTOGRAPHIC FILMS | IONISING GAS CHAMBERS | EXCITING SOLID STATE |
|---|---|---|---|
| measuring | film badge | free-air<br>thimble<br>pocket<br>Geiger | semiconductor<br>thermoluminescent<br>scintillation counters |
| imaging | radiograph | CT scanner | fluoroscopes<br>rectilinear scanner<br>gamma camera<br>CT scanner |

## LEARNING OBJECTIVES

*When you have studied this chapter you should be able to:*

① explain how the physical effects of photography, ionisation and excitation can be used to detect radiation;

② describe common measuring and imaging devices which use these effects;

③ identify detectors which are particularly suitable for measuring and/or imaging each of the radiations X-, $\alpha$, $\beta$ and $\gamma$ for special applications;

④ give an outline of the use of computers in image processing, with particular reference to tomography and digital techniques.

# PHOTOGRAPHIC METHODS

## The radiograph

The remarkable properties of ordinary silver bromide photographic emulsion, make it the most common method of recording images from radiation. It consists of gelatin in which are suspended minute crystals of silver bromide. In X-ray use these have an average linear size of 1 mm. This is supported on a sheet of cellulose acetate, which gives it the name film. In use this is always surrounded by a fluorescent screen, the intensifying screen, which is described in Section 9.3. The X-rays, or other radiation, are absorbed by this screen, and visible and ultraviolet light is produced. This light is absorbed by the silver bromide; when a certain number of quanta have been absorbed, one AgBr molecule is ionised. The silver ion, $Ag^+$, forms a latent image centre which can then be developed into a black silver particle by chemical processing. The density of these black particles is dependent on the intensity of the incident quanta, which therefore produce an image in shades of grey. The processed film is called a *negative* because the darkest parts are those most exposed to the radiation. The processed film is usually viewed in this form, with the use of an illuminated screen. Radiographs are thus similar to slides rather than the prints which most amateur photographers view.

### INVESTIGATION

### Film detection

Use dental X-ray film to detect the radiations from sources available to you in your laboratory.

Two types of film are generally available; one contains the developer and the fix for processing within the packet (like polaroid film). The other which is much less expensive, can easily be processed in the laboratory in dim light, as it incorporates a yellow filter. One side of the film is shielded with lead, so ensure that you use the other!

For sources try the sealed α, β, and γ; the most active of these will require an exposure time of about ten minutes. Other naturally occurring sources can be used, but the exposure times will need to be lengthened accordingly. Try taking a radiograph of a metal object, and try out the effect of filters of plastic and thin aluminium, such as are used in the film badge monitor, described in the next paragraph.

**FIG 12.1** The film badge.

## The film badge

This is the most common type of personal dosimeter. It consists simply of a piece of photographic film in a special holder. It is pinned to the clothing for a period of 1–4 weeks and then processed under standard conditions and the film blackening measured. The film is similar to a dental X-ray film but is a double emulsion type, having a sensitive (fast) emulsion on one side and a slow emulsion on the other.

This enables a wide range of doses to be measured. The holder contains a number of filters, which enable different types of radiation to be identified (Figure 12.1).

The beta dose is measured by comparing the blackening through the open window with that through the plastic filters. These enable the energies of beta to be estimated. The metal filters absorb beta and permit differentiation between photons of different energies. Dural, being an alloy mainly of aluminium, absorbs only low-voltage X-rays, whereas the lead alloys will attenuate all energies. Thermal neutrons (from a reactor), interact with the cadmium alloy, producing gamma radiation and additional exposure.

The fast emulsion enables doses in the range 50 µSv to about 50 mSv to be measured; for higher doses this emulsion is stripped from the film and the slow emulsion then allows measurements of up to 10 Sv. The accuracy is only about 10 to 20 per cent, but this is sufficient for its normal role of low-level monitoring. The indium panel is there to detect higher than 10 mSv doses of thermal neutrons. These doses can then be measured more accurately with a Geiger–Muller tube.

Film badges are cheap and easy to use; they require no maintenance and can be sent away for processing. They provide a permanent record of exposure from a wide range of radiations. The limitations are that the film is sensitive to changes in ambient temperature and humidity, and the results are known only later, which may make it difficult to determine the origin of any unexpected exposure.

---

### QUESTIONS

**12.1** Explain briefly how the developed film shows patterns of blackening as a result of exposure to:
   (a) low energy β radiation only,
   (b) high energy X-radiation,
   (c) direct contamination by a drop of liquid, emitting γ radiation.

**12.2** (a) Why is it particularly important to measure the neutron dose accurately?
   (b) Why will the film badge not measure α radiation, and why is this omission unimportant?

---

## 12.2 IONISATION METHODS – GAS DETECTORS

The result of radiation interacting with a gas is commonly the production of ions, positively and negatively charged particles. These can be collected at electrodes if a potential difference is set up across the ionised material. If the applied voltage is low, many of these ions will recombine. As it increases, a point will be reached when all those produced will be collected. This is known as saturation, and the chamber is said to be operating in the 'ionisation region'. The electrical current resulting from this flow of ions is extremely low (about $10^{-12}$ A), but an electronic circuit known as a d.c. amplifier can be used to measure it. This current represents a mean value of the interaction of many charged particle or photon radiations. This type of device is called an **ionisation chamber**, and is shown schematically in Figure 12.2. Chambers are designed for a range of applications. Significant parameters are the gas with which it is filled, the material of the electrodes and the potential at which it is operated. Those used for health physics monitoring are usually air filled and made of low Z materials.

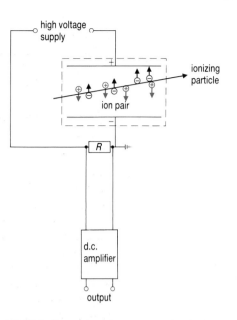

**FIG 12.2** Operation of the ionisation chamber.

# Free-air chamber

The major advantage of the ionisation chamber over other detection methods is its accuracy (~0.1 per cent at best). The free-air chamber is designed to make use of this. It is a device for setting the primary standard in the direct measurement of exposure. It is too cumbersome for medical use but is used in, for example, the National Physical Laboratory (UK), to calibrate other instruments. The design of the device is similar to that shown in Figure 12.2, with careful attention being paid to the dimensions and accuracy of all parts of its operation. For example, the radiation is allowed in as a narrow, collimated beam, approximately parallel to the electrode plates; these plates are constructed to produce a uniform field; the charge is collected from the ionisation of a known volume of air; and it is measured with an electrometer. All these features improve the accuracy of the performance of this instrument.

**FIG 12.3** Thimble chamber **(a)** as exposure meter **(b)** as dosimeter.

# Thimble chamber

This is a small version of the ionisation chamber, suitable for clinical use. The volume is usually less than 1 cm³, and it accepts radiation from a wide range of directions. It is used to monitor exposure or exposure rate, at a point in or on a patient, or simulated patient (called a phantom). It is shown in Figure 12.3, as a small volume of air surrounded by a solid wall which is made of a material with the same mass absorption coefficient as air. This is known as an air-equivalent wall; the material is usually a graphite-plastic mixture, and it forms one electrode of the chamber. The other is a thin aluminium rod in the centre, insulated from the wall by a material such as amber.

The wall is made thick enough not to allow any secondary particles from outside to penetrate it. The ionisation measured is then mainly due to that occurring in the 'air' wall with a small contribution from the air inside. It is operated in the saturated condition, so that all the ionisation charge is collected. This may be measured as a voltage across a capacitor,

$$Q = CV \quad \text{or} \quad V = Q/C$$

and the result is a measure of the total exposure in C kg$^{-1}$. This instrument, an integrating dosimeter, or exposure meter, is shown in Figure 12.3(a). The alternative is to measure the voltage across a resistor $R$ (typically $10^{12}$ $\Omega$ ),

$$Q = It = tV/R$$

or,

$$V = QR/t$$

giving a reading of the exposure rate, C kg$^-$ s$^{-1}$. This instrument, a dosimeter, is shown in Figure 12.3(b).

The thickness of the wall is critical, for accurate readings, and dependent on the energy of the radiation to be measured. If the thimble is to be used to measure radiations of a higher energy than that for which it was designed, the wall can be thickened by the use of a 'build-up cap' such as a sheath of about 3 mm perspex. For lower energies a thinner walled chamber is required.

## Pocket ionisation chamber

For monitoring personal doses it is inconvenient to have the cable connection to the electrometer. This device avoids the problem by using a **capacitor** (or condenser) **chamber**. In this the thimble is adapted to allow the chamber to be charged by contact with the central electrode. Attached to this is a quartz fibre which can move over a scale. In the fully charged condition it reads zero. When the chamber is exposed to radiation there is a gradual discharge as the ions give up their charge, and the fibre moves across the scale. This is calibrated directly in C kg$^{-1}$ or Gy, since these both are proportional to the loss of charge, and

$$Q = C\Delta V$$

where $C$ is the capacitance of the chamber and $\Delta V$ is the voltage change causing the movement of the quartz fibre.

Because of the small size of the movement, the pocket chamber incorporates a built-in microscope (Figure 12.4). The instrument is a convenient method of obtaining a direct reading (usually the range is 0 to 2 $\mu$Gy). It is accurate and independent of energy of the radiation over a wide range. Its disadvantages compared to the film badge, are that it is expensive, provides no permanent record and gives no information as to the type of radiation measured.

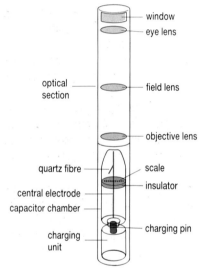

window
eye lens

optical section — field lens

objective lens

quartz fibre — scale

central electrode — insulator
capacitor chamber

charging unit — charging pin

**FIG 12.4** Pocket ionisation chamber.

## Counters

As the voltage applied across the ionisation chamber increases, the charge or current produced also increases as shown in Figure 12.5. This is because the original ions are given sufficient energy to cause secondary ionisation of the gas molecules by colliding with them on the way to the electrodes. This avalanche or 'gas amplification' effect can increase the current greatly and make it possible to detect individual ionising particles as pulses of current. Such an instrument is called a counter. At voltages just above saturation the pulses are proportional to the energy of the original ionising particle. Operating in this proportional region it is possible to identify the original nuclide that produced the ionisation, with the use of a multichannel pulse analyser, which can record both the number of pulses and the total charge of the pulse. Further increases in the applied voltage produce a greater avalanche, such that a constant pulse size is obtained, whatever the energy of the original radiation. This simplifies the electrical circuitry needed, but the type of radiation cannot now be identified. This is called the Geiger region and the instrument that operates in this way is the **Geiger (or Geiger–Muller) counter**.

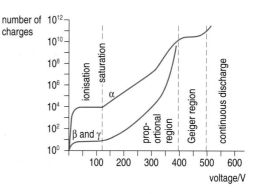

**FIG 12.5** Effect of increasing voltage on a gas ionisation chamber.

# The Geiger counter

You may be familiar with the structure and operation of the Geiger counter from your investigations of the properties of ionising radiations. You may have heard that these are expensive because they are almost exclusively used nowadays in physics education! The counter does have some applications in medicine because of its ability to measure very low activities. It also has the advantage of producing a stable, reliable output with no need for amplification, so the device can be small and portable. Its major disadvantage is the 'dead' time during which it cannot register a pulse since it is recovering from the effects of the previous one. This is about 1 ms, and so puts a severe limit on the maximum activity that can be accurately measured.

It is used in medicine to measure β radiation, including phosphorus-32 inside the body, by means of a special miniature counter, about 20 mm by 2 mm in diameter. Because it is rather insensitive to γ radiation it can be used to count charged particles in the presence of γ. It is also used in the monitoring of equipment to detect contamination.

The structure of the counter is in principle like other ionisation chambers. The gas is a mixture of argon and a 'quenching agent', of a halogen gas to reduce the dead time. The central anode is tungsten, and the body is metal or glass with an inner conductive coating, to make the cathode. The operating voltage is usually between 500 and 1000 V (Figure 12.6).

**FIG 12.6** Geiger–Muller tube.

---

**QUESTION**

**12.3** With reference to the preceding text, make notes on the following aspects of the Geiger counter.
(a) Explain how it produces a pulse of current from an avalanche of charge, with the aid of a labelled diagram.

**(b)** Explain the cause of the 'dead time'.

**(c)** Estimate the maximum activity which can be measured, from the value of the dead time quoted above.

**(d)** Why is background radiation a particular source of error in Geiger counts, and how may this be accounted for?

# EXCITATION METHODS – SOLID STATE DETECTORS

## 12.3

The effect of radiation on many crystalline substances is to raise the energy of electrons in it. The highest level in which the electrons normally exist is the valence band, above which there is a forbidden band. The transfer of energy may raise the electron through this to the conduction band or the exciton band, or the presence of impurities may give rise to the electron being trapped in the forbidden band (Figure 12.7). In the conduction band electrons are free to migrate, when a potential difference is applied. In the exciton and forbidden bands they are not, but the trapped energy can be released to provide a measurable signal. In the exciton band this occurs spontaneously; in the forbidden band it must be stimulated. Each of these effects gives rise to detectors, which will now be described.

**FIG 12.7** Electrons and energy: conduction, excitation and trapping.

## Semiconductor detectors

In substances, such as cadmium sulphide, silicon and germanium, the energy of the radiation excites an electron from the filled valence levels to the conduction levels. This provides a pair comprising a conduction electron, and an electron vacancy which is called a positive hole. The energy to produce this is about 1eV, which is low, so the potential sensitivity of the detector is high. If a potential difference is maintained across the semiconductor the charge carriers will move and a current pulse will be observed. The devices have been used in place of Geiger counters, though they are expensive and rather insensitive to $\gamma$ radiation.

## Thermoluminescent dosimeters (TLDs)

These devices are replacing film badges for monitoring personnel, and are also used for monitoring patients' doses during therapy and calibrating radiation sources in hospitals. The principle of the TLD is electron trapping. Certain materials such as lithium fluoride (LiF), and calcium fluoride (CaF$_2$), can, when exposed to radiation store part of the energy in electrons trapped in the isolated level resulting from impurities in the material. They cannot migrate and conduct, but if the crystal is heated they can gain thermal energy which then enables them to lose all their additional energy as light. The amount of light emitted is a measure of the number of trapped electrons, and hence the original exposure.

Substances with the ability to store and subsequently release light energy are called **phosphors**. LiF is the most commonly used in TLDs as it has the following features:

■ it can detect γ, β, X- radiations and thermal neutrons,

■ it responds to a very wide range of total energy and rate of energy deposition,

■ it has a similar mass absorption coefficient to tissue and air, so that dose estimates can readily be made,

■ it is sensitive and gives reliable results,

■ the light is released on heating to a suitable temperature (~200 °C).

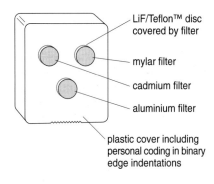

LiF/Teflon™ disc covered by filter

mylar filter

cadmium filter

aluminium filter

plastic cover including personal coding in binary edge indentations

**FIG 12.8** Thermoluminescent dosimeter.

The phosphor is usually incorporated into the plastic Teflon, and produced as a disc. This can be attached, in a light-proof plastic container, to the body or clothing. Fingers and hands may be measured directly in this way, for example by taping a plastic pouch to the appropriate part – that which has been contaminated. Alternatively the discs are incorporated into a badge, worn over a period of time like a film badge (Figure 12.8). Different filters in front of the thermoluminescent discs make it possible to measure the effective dose to different parts of the body, for example, mylar for skin and aluminium for the gonad dose.

TLDs are inexpensive, requiring only small quantities of LiF, which can be re-used, and they can be made very small to measure exposure within the body at sites such as tumours. The necessary heater and light emission analysis unit is expensive however; to be economic their use needs to be widespread.

## Fluoroscopes

A fluorescent substance is one in which the energy of radiation moves electrons into the exciton band. They then immediately re-emit this absorbed energy of radiation, as light. Dynamic imaging of the body with X-rays was carried out for many years by use of a glass screen coated with zinc-cadmium sulphide, the fluoroscope. Direct viewing is no longer practised, because of the high exposure to X-rays required. Modern fluoroscopy uses an image intensifier tube as described in Chapter 9. The fluorescent screen is backed with a photocathode which changes the light into electrons. These are then accelerated, and focused on to another fluorescent screen, to produce a gain in brightness of about 5000. The output can then be viewed directly, photographed, or videotaped, using a special camera tube. Direct viewing is rarely used; electronic processing of the video image offers computer image-enhancement possibilities as well as recording facilities.

## Scintillation counters

These operate on the same principle as fluoroscopy; the emission of light as an electron returns to the valence level. In this case the purpose is to count the emissions or scintillations, and therefore measure the exposure.

There are a number of **scintillator** materials in use including those in solution, and large organic molecular crystals, which are used for detecting β particles. Those most widely used are simple ionic crystals such as zinc sulphide (to detect α), lithium iodide (for thermal neutrons) and sodium iodide (for γ).

The most common scintillator is sodium iodide, activated with about 0.5 per cent thallium iodide, so it is usually abbreviated as NaI(TI). The substance is hygroscopic, so it has to be kept in sealed containers. These are usually made of aluminium, with a glass or quartz window which is coupled to the photomultiplier tube with silicone oil or grease.

It has the advantages of

■ high efficiency in converting about 10 per cent of the incident energy into light,

■ rapid scintillation,

■ excellent γ detection ability, due to its high density, and relatively high proton number.

The effect of the light emission can be magnified with a **photomultiplier tube**, which is a development of the image intensifier discussed earlier. Light from the scintillator ejects electrons from a photocathode. This is a thin layer of semiconductor material coating the inner surface of the tube. These photoelectrons are then accelerated and multiplied by a series of electrodes called dynodes, held at increasingly positive potentials along the tube. The dynodes are made from an alloy such as beryllium-copper, which causes several secondary electrons to be emitted by the energetic incident electron. This process is repeated several times, causing multiplication each time. The electrons are collected by the anode and output through a further amplifier. The complete device is called a scintillation counter (Figure 12.9), and it is able to amplify the original light output by a factor of between $10^5$ and $10^8$, but strictly in proportion to the original energy. By using a multichannel pulse analyser it is possible to obtain the energy spectrum of the incident radiation.

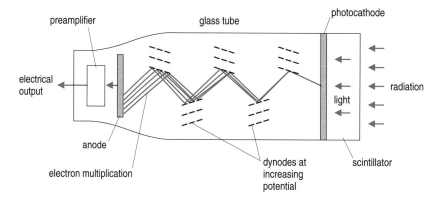

**FIG 12.9** Photomultiplier tube.

The scintillation counter has widespread use in hospitals, as it can measure all types of radiation commonly encountered, with the selection of a suitable scintillator. It is particularly favoured for the detection of X- and gamma radiations, and can be used to analyse the energy spectrum of γ, and thus identify the source nuclide. It is also used to monitor low intensity tracer levels in investigations using the gamma camera (see next section). The main restrictions on its use are its relative expense and its size.

## QUESTIONS

**12.4** (a) Explain how solid state detectors make use of electron excitation effects to detect radiation.

(b) Describe the physical principles underlying the action of a scintillation detector and photomultiplier, when used as a scintillation counter.

**12.5** Table 12.2 summarises the most commonly used detectors for various purposes. List the factors which are important in choosing a detector, and account for the particular choices listed.

**TABLE 12.2** Most common uses of detectors

| TYPE/SITUATION | FILM BADGE | TLD | THIMBLE CHAMBER | PIC | GEIGER COUNTER | SCINTILLATION COUNTER |
|---|---|---|---|---|---|---|
| low energy X-rays | | | ✓ | | | |
| high energy X-rays | | | | | | ✓ |
| α particles | | | | | ✓ | |
| low energy ß | | | | | ✓ | ✓ |
| high energy ß | | | | | ✓ | |
| γ rays | | | | | | ✓ |
| personnel (general) | ✓ | ✓ | | | | |
| personnel (protected areas) | | | | ✓ | | |
| low activity | | | | | ✓ | |
| local and internal body sites | | ✓ | | | | |

# RADIOISOTOPE IMAGING – SCANNERS AND CAMERAS

**12.4**

**T**he process of creating a two dimensional image from the emissions of radionuclides within the body, is quite different from that for X-rays. Earlier techniques included the laborious moving of a Geiger counter backwards and forwards across the body. This process of scanning can be automated and the scintillation counter is used as this is a more suitable detector for gamma radiation.

## The rectilinear scanner

This imaging device was developed in 1950, using these principles, as shown in Figure 12.10(a). The detector is collimated so that it receives radiation from a particular place, when in a particular position. Interference from emissions from different depths can be reduced by a special 'septa', a lead collimator which is 'focused' on a fixed plane (Figure 12.10(b)). The movement of the detector is replicated by a light source which produces an exposure on film to match the intensity of the radiation. Such scanners have been widely used in hospitals, but for most applications they are now superseded.

(a)     (b)

**FIG 12.10** Rectilinear scanner **(a)** scanning method **(b)** focusing method.

# The gamma camera

This was invented in 1957 by Anger, which gives it its alternative name. The purpose is the same as the rectilinear scanner, but the detector is designed so that no movement is needed. Essentially it consists of a large disc of sodium iodide, typically 40 cm in diameter and 1 cm thick (Figure 12.11(a)).

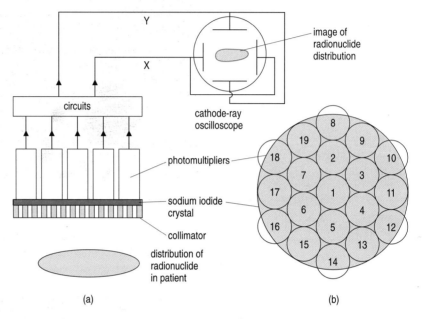

(a)

(b)

**FIG 12.11** Gamma camera **(a)** scheme of operation **(b)** arrangement of photomultiplier tubes.

To this are attached an array of photomultiplier tubes. These are arranged in a geometric pattern. Figure 12.11(b) shows a common arrangement of 19; more advanced machines use up to 75. These tubes will receive signals from each flash of light in the crystal. Because they are different distances from it, the intensity will differ. The result is 19 output pulses of varying size, from which, with the help of some clever electronics, called position decoding circuits, the position of the light flash can be deduced. This is then converted to signals representing its $x$ and $y$ coordinates so that it may be displayed on a cathode-ray screen, and a Z coordinate for its brightness.

A lead collimator is used as in the rectilinear scanner, to restrict the directions from which the gamma radiation is received. Usually it is a circular slab of lead pierced with many holes perpendicular to the face. This ensures that only those gamma rays are received which are emitted at right angles to the crystal and the image is not degraded by those received from other directions. Sometimes collimators are used with non-parallel holes, for example to reduce the field of view, or with a single 'pinhole' for small organs such as the eye or thyroid.

The isotope most commonly used is $^{99}$Tc$^{m}$ with a γ photon energy of 140 keV, an alternative is $^{123}$I with a photon energy of 160 keV. A change in the collimator thickness allows for the energy difference between the two gamma emissions.

The gamma camera image is formed, in two to three minutes, of up to one million small dots, the frequency of which is proportional to the region with the greatest concentration of the radionuclide (Figure 12.12). The resolution of the image is about 3 mm, limited by the large size of the photomultiplier tubes. The further development of solid state detectors is expected to improve this. Computer processing permits a range of additional processing to be carried out. Examples of their wide applications in diagnostic imaging were given in Section 10.5.

**FIG 12.12** Gamma camera image.

Variations on the standard camera include a larger crystal for a larger field of view (Figure 12.13), a mechanical scanning method to cover the whole body, and a rotating detector to produce tomographic images.

**FIG 12.13** A large field of view gamma camera.

## QUESTIONS

**12.6** **(a)** Draw a diagram showing the essential components of a γ camera. Your diagram should indicate the position of the patient and the origin of the γ photons. Name a suitable isotope for use in this procedure.

State three factors to be considered when choosing a radioisotope suitable for use in diagnosis.

*(London Jan 1996)*

> **12.7** **(a)** Explain how a gamma camera can be used to obtain an image of a particular part of the body.
>
> **(b)** Compare such an image with that obtained from a diagnostic X-ray.

## 12.5  TOMOGRAPHY – CT SCANNING

### Tomography

This is a technique for producing an image of a slice of the object, taking its name from the Greek *tomos*, a section. At its simplest it is a way of arranging to have one plane produce a clear image, regions above and below being blurred. Since X-rays cannot be focused by lenses, this is arranged by moving the X-ray tube and the film cassette. The two are connected mechanically so that, as they move, the image of the chosen plane stays in the same place on the film. In Figure 12.14 the image of A stays at A'. Objects not in this plane, such as B are imaged in different positions as the unit moves, from B' to B'' in Figure 12.14. This simple translation movement can introduce artefacts (mistakes) in the images, so more complex cycloidal or elliptical paths are often used.

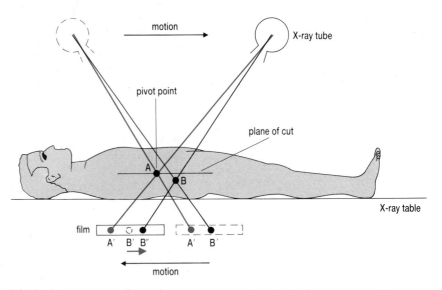

**FIG 12.14** Linear tomography: as the camera moves to the right, the film moves to the left, and the image of B moves from B' to B'', while the image of A stays at A'.

Godfrey Hounsfield was born in England in 1919 and educated in Newark and London as an electrical engineer. He joined the medical systems section of the firm EMI in 1951 and his long career in medical research and engineering culminated in his invention of the computed axial tomography scanner, often called the EMI scanner. This earned him the Nobel prize for medicine in 1978.

In this example the plane imaged is a longitudinal one. In **axial tomography** the film and tube must rotate around the body to image a cross-section. These techniques were greatly enhanced in 1971 when Hounsfield developed a computer method of analysing and displaying the images. This **computed tomography** enables detailed pictures to be produced of particular planes of the body. It is often referred to as **CT** or **CAT** (computed axial tomography) **scanning**. It is a method of measuring the amount of X-ray absorption at a specific place, so that the information can be stored digitally in computer memory, recalled and processed to give a selection of reconstructed images.

Similar methods can be applied to radionuclide emissions where the technique is called emission computed tomography or **ECT**. The method which uses the rotating gamma camera to detect photons in the way described above, is called single photon ECT, or SPECT. A second method is more complex. It uses two rotating detectors to detect the annihilation radiation emitted from the patient who has been injected with a positron emitting radionuclide. This is known, not surprisingly, as positron emission tomography, or **PET**. Computed tomography is also used with a completely different modality, magnetic resonance imaging, or **MRI**. This will be considered in the next chapter.

## *Principles of computed tomography*

The task is to make an image of a section of the body, from measurements made around its axis (Figure 12.15(a)). The plane is divided into small volume units called voxels (it is assumed that the slice has a finite thickness). The intensity from a given voxel, called a pixel, is reconstructed from the measurements made along a series of strips across the whole plane. The principle may be illustrated by back projection which is illustrated in Figure 12.15(b), though more sophisticated mathematical

**FIG 12.15** Computed tomography **(a)** arrangement of equipment **(b)** summing pixels horizontally **(c)** turn through 45° and add new intensities **(d)** repeat **(e)** repeat **(f)** subtraction of background and reconstruction.

methods are used in practice. The total intensity measured along any strip is simply allocated to each pixel in that row. For example, viewing the sample array of four in (b) from the right, gives totals of 7 and 9. A second view at 45° to this adds 1, 11 and 4. These are added in (c) to give new totals. The procedure is repeated in (d) and (e), to give totals of 25, 28, 19 and 40. The final step is to remove a background of 16 from each and divide by 3 because of the duplication of view. The original values of 3, 4, 1 and 8 reappear, as if by magic! As we said this is an illustration of the principle, in practice the computer takes many more steps and uses rather more complex formulas.

## CT equipment

The **X-ray source** must be highly collimated to produce the precise paths required, typically each strip of image is only a few millimetres wide. This means most of the radiation energy is wasted. The tube voltages and exposures are greater than for conventional radiography, typical values being 130 kV and 300 mA s$^{-1}$. Therefore a great deal of heat has to be dissipated from the anode. Because of the tube movement during scanning, they tend to fail after a few months in service and replacements are a large expense in the running of a CT department.

**Filtration** This is very important, to ensure the beam is homogeneous and that the patient skin dose is reduced. In the earliest models the patient was surrounded by a water bag to absorb stray unattenuated beams. Present models can avoid this problem. An array of several hundred scintillation counters act as the **detector**, mounted so as to receive the beam as the tube rotates (Figure 12.14) or able to match the movement in earlier models. The traditional scintillator for X-rays is NaI(Tl). This has some limitations in the scanner and is being replaced by caesium iodide thallium-doped crystals coupled to silicon semiconductor photodiodes. Other models use xenon gas ionisation chambers.

The X-ray tube and detectors are mounted within the scanning **gantry**. This is a frame with an aperture in which the plane of the body to be scanned can be positioned (Figure 12.16). It contains the driving mechanisms and the sequencing sensors. The frame can be set at different angles for alternative sections. The **couch** for the patient is motorised to allow accurate positioning of the body in the scan plane. A single scan takes between 2 and 8 seconds.

**FIG 12.16** CT scanner in action with positioning the patient in the gantry aperture.

The collection, reconstruction and display of the scan data requires a powerful **computer** and sophisticated peripherals and **software**. The image reconstruction of each pixel intensity value is the result of about one million computations. The image is displayed on a TV screen to produce a picture of the slice scanned with variable contrast and brightness. It can be stored in the computer memory and used to construct a slice in a different plane in combination with all the other pixel data.

## 12.6 DIGITAL SUBTRACTION

### *Subtraction principles*

**FIG 12.17** Digital subtraction image of the blood supply to the kidney.

The technique of 'subtracting the background' is often used to increase the contrast of an image, so that more detail can be detected. For example blood vessels, being soft tissue, are often obscured in radiographs by the greater contrast of bone. A contrast medium can be injected into the bloodstream so that it changes the intensity of the image. If two pictures are taken before and after the injection, only the blood vessel intensity will vary. The contrast of one picture is now reversed, by making a negative, i.e. dark for light. When the two pictures are combined together the common parts such as the bone will be reduced in intensity (one dark and one light) whereas the blood will be increased in intensity (both light, or both dark). This has been a well established 'analogue' technique in X-radiography. The advent of computing power has enabled this to be developed. In this respect it is similar to computed tomography.

Angiography, the displaying of the blood vessels is at present the most important application of digital subtraction (Figure 12.17). However, the increased availability and lower cost of computing power, is certain to make digitising of radiological images more widespread in the near future.

**QUESTION**

**12.8** Read the article on digital subtraction angiography in the box below, and answer these questions on it.
(a) Why is it particularly useful to image the blood vessels?
(b) What is a catheter?
(c) Why does the fluorescent dye show up under X-rays and how is this use different from the contrast effects from a barium meal?
(d) What is misleading about the fourth paragraph?

**CASE STUDY**

### *New X-ray equipment for safer treatment*

People who once would have bled to death can be cured by doctors manipulating a tube inside their arteries. The doctors are guided by watching an X-ray image replayed on a television screen.

Although the technique to show doctors what is happening in blood vessels has been available for several years, new X-ray equipment is making the process safer. It is also extending the range of complaints that can be treated.

With the technique, called digital subtraction angiography, doctors can pinpoint the position of tumours and blocked or damaged blood vessels.

To see the blood vessels, the doctor first takes an ordinary X-ray, which does not show soft tissue. The image is converted to digital information. The doctor then inserts a tube, or catheter, into one of the patient's veins or arteries, helped by the X-ray image on the TV and a knowledge of anatomy.

When the catheter is in position, a fluorescent dye which shows up under X-rays is pumped into the patient, and more X-rays are taken and stored as digital information recorded by the conventional X-ray. From the image recorded with fluorescent dye in the blood vessels, the doctor obtains a picture of the blood vessels. And from the picture a diagnosis is possible.

The usefulness of the X-ray equipment and digital subtraction angiography does not stop with diagnosis. For example, orthopaedic surgeons carried out a biopsy on the fibula of a patient and damaged an artery. Guided by the X-ray image, doctors were able to insert steel coils in the artery to block it off and stop her bleeding to death.

*New Scientist 31.1.1985*

## SUMMARY ASSIGNMENTS

**12.9** Figure. 12.18 shows the location of a series of filters on a film badge. The photographic film is wrapped in light-proof paper.

**FIG 12.18**

Figure. 12.19 shows a developed film which had been encased in the film badge of Figure 12.18 and worn by a worker in a radioactive environment.

**FIG 12.19**

Deduce the type of radiation to which the worker has been exposed. Explain your reasoning with reference to the penetrative properties of the following radiations:

α-particles; β-particles; γ-radiation; and thermal neutrons.

*(OCR Mar 1999)*

**12.10** Figure 12.20 shows the main components of a gamma camera.

**FIG 12.20**

(a) For each of the components labelled on the diagram, give its name and state its function in the detection of radiation.

(b) Give **one** example of the use of a gamma camera in medical diagnosis.

(c) Describe the procedure which would be followed to obtain an image in the example you have given.

(d) How does the information obtained from such an image differ from that obtained from a diagnostic X-ray image?

*(NEAB Mar 1997)*

**12.11** Write an essay on the detection of gamma radiation. Your account should include

(a) an outline of the structure and mode of operation of a scintillation counter,

(b) an outline of the structure and use of a gamma camera,

(c) a comparison between the method of production of images by a gamma camera and a rectilinear scanner.

*(OCR June 1999)*

# 13 MAGNETIC RESONANCE IMAGING

**FIG 13.1** NMR scanner generating image of head.

**M**agnetic resonance imaging (MRI), is a completely new method of imaging, or modality, which has become an important diagnostic tool in the last ten years. It is based on the phenomenon of **nuclear magnetic resonance (NMR)**, a process in which protons interact with a strong magnetic field and radio waves to generate electrical pulses that can be digitally processed. The medical application of nuclear magnetic resonance began in the 1950s, but the first images of live patients were not produced until the late 1970s. The images produced are in many ways similar to those of the X-ray CT scanner, but the technique is particularly good for soft tissue imaging (Figure 13.1). Although the equipment is complex and expensive, it has the great advantage over X-ray and other ionising radiation methods – there is no radiation hazard.

## LEARNING OBJECTIVES

*When you have studied this chapter you should be able to:*

① explain the physical processes involved in nuclear magnetic resonance;

② describe how the signals generated by the NMR are used in imaging;

③ give examples of the uses of MR imaging, and their advantages and disadvantages, compared to other imaging modalities.

# NUCLEAR MAGNETIC RESONANCE

**T**he complexity of the equipment is a direct result of the complexity of the phenomenon. We will take it step by step. Each of the three steps involves complex physics. You will probably need to read this more than once to take it in!

## Nuclear processes

The nuclei of atoms which have an unequal number of neutrons and protons, **spin**. Because the nuclei are charged, and moving charges produce a magnetic field, they act like tiny magnets.

(a)

(b)

## Magnetic processes

In a magnetic field the nuclei will align themselves, just like iron filings do on a larger scale. This alignment is not exact however, and the nuclei rotate around the direction of the field, while spinning. This motion is called **precessing**, and it is exactly the same as the motion of a top spinning in gravity (Figure 13.2). The frequency of the precession depends on the magnetic field and on the nature of the nucleus. It is called the **Larmor frequency** and is found to lie in the radio frequency (RF) part of the electromagnetic spectrum.

**FIG 13.2** Ways of spinning
**(a)** on its own axis
**(b)** spinning and precessing around the vertical axis.

## Resonance

If a pulse of radiation of radio frequency is applied to the body, those nuclei with the matching frequency will resonate, and absorb energy. When the pulse ends the nuclei return to their equilibrium state with the emission of RF radiation. This occurs over a short period of time, called the **relaxation time**. This signal can be detected and forms the basis of MR image formation.

These steps are illustrated schematically in Figure 13.3.

atoms in body have NUCLEAR spins

strong MAGNETIC field aligns spins

pulse of rf field causes RESONANCE of nuclei

$B_0$

**FIG 13.3** NMR schematic of processes.

Examples of nuclei able to resonate in this way are hydrogen, phosphorus and carbon. Because of its abundance, in body fluids in particular, hydrogen is the nucleus used in imaging.

**INVESTIGATION**

### A simple model of magnetic precession

You can show the process of precession in a magnetic field with a plotting compass and two bar magnets.

In Fig 13.4(a) the compass needle is aligned with the field B of the first bar magnet.

When a second bar magnet is introduced briefly so that it moves perpendicularly as shown in Fig 13.4(b), the needle oscillates about the direction B. This is a two-dimensional model of how the protons precess.

Try moving the compass further from the first magnet and repeating the movement of the second. Changes in the frequency of the oscillation demonstrate the dependence of the Larmor frequency on magnetic field strength.

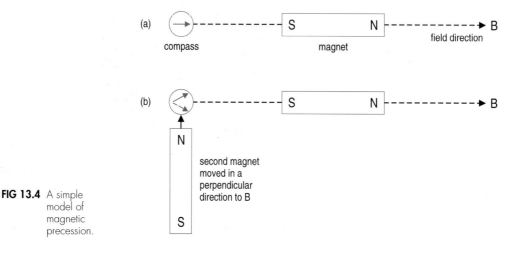

**FIG 13.4** A simple model of magnetic precession.

# THE MAGNETIC RESONANCE SCANNER

**T**he components of the MR scanner are shown in Figure 13.5.

## The main magnet

This needs to have a field strength of up to 2 tesla which is uniform and constant, over a volume large enough to accommodate a patient (Figure 13.6). The options are to use permanent magnets or electromagnets made either of copper or superconducting material.

**Permanent magnets** These are made of alnico (an aluminium-nickel-cobalt alloy), are heavy so only low field strengths (up to 0.2T) are available, but they can have large apertures, which is less claustrophobic for the patient.

**Electromagnets** Large electromagnets can be used to produce field strengths of up to 0.5T. If conventional copper coils are used, a heavy current is required to generate the high field. This produces a substantial heating effect (as high as 150 kW), due to the resistance of the coil. This is removed by water circulating through copper tubes of which the coil is made. The materials of the magnet are relatively inexpensive, but the energy dissipation means that the operating costs

Superconductivity was discovered by the Dutch physicist Heike Onnes in 1911. Three years earlier he had succeeded in liquefying Helium at 4 K, which was required to reach the low temperatures at which the phenomenon occurs. This behaviour is due to the crystal lattice not absorbing any electrons from the electrons which pass through it, carrying the current.

are high. The resistive heating effect of electromagnets can be avoided by making use of the property of superconductivity. This occurs in many materials at very low temperature. The niobium–titanium alloy used to make the coils requires a temperature below 10 K, so it needs to be immersed in liquid helium, the **cryogen**. The loss by evaporation of this expensive substance is reduced by surrounding it by liquid nitrogen, and surrounding the tanks of both by a vacuum. The replenishment of these cryogens is the major running cost of this equipment, though it is lower than the energy costs of the conventional electromagnet. As there is virtually no resistance in the coils, magnetic field strengths of up to 2T can be achieved with little power use. The equipment is costly to build and takes some time to set up, because of the cooling required.

**FIG 13.5** Components of an MR Scanner.

**FIG 13.6** Structure of an
MR scanner.

## Field gradients

The equipment must include a method of creating a gradient in the static magnetic field, so that the signals detected from the nuclei can be located in position. There are usually three electromagnets to produce about $0.01\,\mathrm{Tm^{-1}}$ in three perpendicular directions.

> ### QUESTION
>
> **13.1**   The strength of a magnetic field, the flux density is given by
>
> $$B = \mu_0 nl$$
>
> Where $\mu_0$ is the permeability of free space, $n$ the number of turns per unit length and $I$ the current.
>
> How would an electromagnet be designed to have a magnetic field with a linear gradient of flux density?

## Radio frequency transmitters and receivers

Radio frequency pulses are supplied by yet more coils, tuned to the resonant frequency. RF pulses are transmitted to the patient through an RF monitor to ensure the correct frequency, amplitude, phase, bandwidth and repetition rate to produce the required contrast from the tissue. Because there is a time interval between the transmission and reception of the signals, the same coils are often used to detect the MR signal. These coils are then acting as transceivers.

## Display and storage

The MR signals are digitised and processed by computer. They can then be stored and a range of displays produced from the 3-D scan data. Often the images will be presented slice by slice in a particular plane, or 3-D images can be constructed. The computer processing is now sufficiently fast for real-time imaging to be possible, if required.

## Site requirements

Rooms of suitable size and reinforced floor construction are required because of the weight of the magnet and the size of the scanner and computer equipment. Screening from the magnetic field is required; this is usually by enclosing the machine in a metal cage, lining the room with steel, and providing additional shielding magnetic fields to counteract the main fields.

## 13.3 USES OF MRI

The origin of the MR signal is the hydrogen nucleus, so images can be obtained from all tissues of the body. Water retention in local areas for example will result in increased signals so that inflammation, swelling, bleeding and cysts can be detected. MRI is particularly useful in applications which are difficult using other modalities. It was developed originally for neurological investigations, where the detailed 3-D imaging is ideal for detecting tumours and other abnormalities in the brain and spinal cord. Other delicate and complex tissues and organs which have been successfully imaged by MRI include the kidney, liver, pancreas and bladder (Figure 13.7).

**FIG 13.7** NMR image of internal organs.

In twenty years of research and clinical investigations, there have been no reports of any adverse effects from the magnetic or RF fields. The potential dangers are of induced currents in blood flows due to the high and changing magnetic fields, and local heating due to the RF fields. Precautions required include:

- the removal of any magnetic objects from the vicinity of the scanner;
- removal of magnetically sensitive devices such as cardiac pacemakers and hearing aids;
- careful checking if patients have metal implants such as joints, though these are now usually made of non-magnetic materials;
- care if patients have metal dental fillings in which currents could be induced to produce local heating.

There are a number of aspects of the MRI scanning procedure which though not hazardous, are potentially unpleasant for the patient:

- being enclosed in the core of the magnet may induce claustrophobia;
- the rapid switching of the gradient coils is very noisy;
- the patient must remain immobile during quite long examination times.

**Magnetic resonance spectroscopy** uses signals from a range of different nuclei to study body chemistry in-situ (Fig 13.8). It will provide information as well as images, to help answer questions about for example, how anticancer drugs are taken up by, and kill, tumour tissue.

**FIG 13.8** *Magnetic resonance spectroscopy.*

## SUMMARY ASSIGNMENTS

**13.2** The nuclear magnetic resonance effect (NMR) is used for medical imaging. Images are produced from information collected from atoms in the body, particularly hydrogen atoms.

   **(a)** In what sense is the technique *nuclear*?

   **(b)** Explain the advantages of NMR over X-rays as an imaging method.

   **(c)** NMR requires three separate magnetic fields. Explain the chief function of
   - **(i)** the strong uniform field,
   - **(ii)** the field gradient supplied by secondary coils,
   - **(iii)** the short pulse magnetic fields which oscillate at radio frequencies.

   *(London June 1997)*

**13.3** Write an essay on magnetic resonance imaging (MRI) and the use of lasers in endoscopes for diagnostic purposes. Your account should include

   **(a)** an explanation of the process by which an MRI signal is produced,

   **(b)** an explanation of the principles of operation of an endoscope using light to image an internal structure,

   **(c)** **(i)** explanations of why MRI is used in preference for imaging the spine, and endoscopes are used for imaging structures within the stomach,

   **(ii)** a discussion of the relative merits and limitations of each technique, making reference to one further specific example of the use of each.

   *(UCLES June 1997)*

**13.4** **(a)** Ultrasonic scanning depends upon the sonar principle. What is meant by the *sonar principle*?

   A patient has several pieces of glass embedded in his thigh, half-way up from the knee. Explain how an ultrasonic technique can be used to create an image of a cross-section of the leg to determine the positions of the larger pieces of glass.

   What factor determines the size of a piece of glass which is just too small to be detected? Explain your answer.

   If you were free to choose either ultrasonic or X-ray equipment for this task, which would you choose? Give reasons for your choice.

   **(b)** A doctor needs to obtain an image of a horizontal section of a patient's brain. State and explain two advantages and one disadvantage of nuclear magnetic resonance equipment for generating the image.

   *(London Jan 1997)*

# Appendix A: Answers

## Chapter 1

**1.1** See Figure 1.20 (a) and (b)

**(a)** Looking up, **(b)** Standing on tiptoe.

**FIG 1.20**

**1.2** Biceps force = 9036 N    Elbow reaction = 7756 N
(assuming each arm supports half the weight at the palm of the hand)

**1.3** Compressive stress = $0.7/2\ W \times \cos 50° = 0.35 \times$
700 N $\times$ 0.6428 = 157.5 N
Shear stress = $0.7/2\ W \times \sin 50° = 187.7$ N

**1.4** From the figures given in the examples in the text and answer to question 1.3:
157.5 : 2385 : 4886 or 1 to 15 to 31

**1.5** **(a)** **(i)** $F = \mu V$ where $\mu$ for shoes is 0.7 and $V =$ body weight $W$. i.e. $F = 0.7\ W$
For walking horizontal force is about $0.2W$, easily opposed by friction.
**(ii)** For a given body weight and ground surface if $\theta$ is smaller $F$ is smaller.
**(b)** We suggest you do a practical investigation on this one!

**1.6** **(a)** see text
**(b)** high initial centre of gravity
**(c)** increasing forward momentum
**(d)** keeping centre of gravity low in relation to body position. It is possible for it to stay lower than the bar, yet for the body to clear the bar.

**1.7** **(a)** About 1 kg per hour
**(b)** **(i)** A large area of the skin is exposed, other methods of loss are not sufficient to produce the rate of cooling required.
**(ii)** The body is almost completely covered, any sweat produced will not evaporate.
**(c)** The amount of energy loss is still low but the other mechanisms are hardly operating.

**1.8** Read off from the graph, the value of BMR in KJ m$^{-2}$ h$^{-1}$ and convert this to watts using values of surface area, from Table 1.2.

**1.9** Example of a day's energy expenditure:
Average value from the range given in Table 1.3 and a moderately active day: 8 h sleeping + 6 h sitting and standing + 9 h moving and $\frac{1}{2}$ h climbing and carrying + 1 h walking and housework + $\frac{1}{2}$ h moderate sports = 14.3 MJ

**1.10** **(a)** 363.06 kJ h$^{-1}$
**(b)** 100.85 W
Panting increases the rate of supply of oxygen required for the higher rate of respiration. Massaging brings oxygenated blood to oxidise the lactic acid in the fatigued muscles.

**1.14** **(a)** **(i)** See Figure 1.21

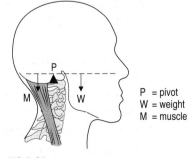

P = pivot
W = weight
M = muscle

**FIG 1.21**

**(ii)** Taking moments about P:
$M \times 0.055 = 75 \times 0.040$
Force of neck muscles is 54.5 N

**1.15** **(a)** **(i)** Refer to Figure 1.22

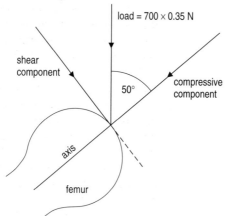

**FIG 1.22**

**(ii)** Compressive component = $700 \times 0.35 \times \cos 50° = 160$ N
Shear component = $700 \times 0.35 \times \sin 50° = 190$ N

**(b)** **(i)** Pressure = $\dfrac{\text{(normal) force}}{\text{area}}$

$$= \frac{0.6 \times 700 \times \cos 40°}{3 \times 10^{-4}}$$

$$= 1.1 \times 10^{6} \text{ N m}^{-2}$$

**(ii)** See Section 1.2

**1.16** **(a)** See section 1.4.

**(b)** From the graph BMR = 162 KJh$^{-1}$ m$^{-2}$

$\therefore$ rate = $\dfrac{162 \times 1.7 \times 1000}{60 \times 60} = 76.5$ W

**(c)** **(i)** $\dfrac{7.69 \times 60 \times 60 \times 24}{16 \times 10^{6}} = 0.413$

**(c)** **(ii)** $\dfrac{16 \times 10^{6} \times 0.587 \times 3}{4 \times 60 \times 60 \times 24} = 81.5$ W

**(c)** **(iii)** Because this answer in **(ii)** averages over 24 hours – part of which has minimal activity.

**1.17** **(a)** $1500/1.8 = 833$ strides
$P = mgh/t$
$P = 833 \times (80 \times 9.8 \times 15 \times 10^{-2}/4.0 \times 60)$
$P = 408$ W

**(b)** See section 1.4

**(c)** **(i)** o/p power = 1600 W
$1600 \times 75/100$ = rate of production
= 1200 W

**(ii)** $40/100 \times 1200$ = 480 W
$480 \times 4 \times 60$ = $1.15 \times 10^{5}$ J
$1.15 \times 10^{5}/2300$ = 50 g

**(d)** See Section 1.4

## Chapter 2

**2.2** There is a greater density of rods.

**2.3** About one minute, depending on the brightness.

**2.4** It stimulates the red and green receptors equally; red + green = yellow.

**2.5** Cyan is the complementary colour to red. The white light from the wall will stimulate all the receptors. If the red cones have been depleted of their photosensitive agents, in the region of the ball's image, then only blue and green will operate. One test would be to try different colours of wall.

**2.8** The refractive index of water is about 1.33, close to that of the cornea, reducing the refractive power from ~ 46 D to ~ 6 D.

**2.10** 10 m, which is probably rather further than you can see. Can you explain the difference? Remember this is only a distinction between two points, to recognise an object, the brain requires more information.

**2.11** **(a)** Diverging, –0.5 m
**(b)** **(i)** –5 D **(ii)** 0.11 m **(iii)** 0.25 m (normal vision).

**2.13** **(a)** Long sight, use lenses of power +3 dioptres
**(b)** Presbyopia, use bifocals, with the lower half of power +2 D and upper half –0.2 D.
**(c)** Astigmatism, use a lens with a horizontally oriented cylindrical correction.
**(d)** Diplopia, possible correction by exercising the rectus muscles to strengthen them.

**2.14** **(a)** Chromatic aberration has caused the eye to be unable to superimpose the two images.
**(b)** The image of the right hand object has fallen on the blind spot of the right eye.
**(c)** Reducing the size of the viewing aperture has increased the depth of field. Reducing it to 1 mm has caused significant diffraction effects.

**2.15** See Section 2.3

**2.16** **(a)** See Section 2.1
**(c)** See Section 2.5
**(i)** 3.2 D **(ii)** positive.

**2.17** See Section 2.4

**2.18** **(a)** See Section 2.1
**(b)** **(i)** $P = 1/f$
$f = 1/59 = 1.7 \times 10^{-2}$ m
**(ii)** $P = 1/v + 1/u$
$P = 1/1.69 \times 10^{-2} + 1/60 \times 10^{-2}$
= 60.8 D
or
$P = 59 + 1/0.60$
= 60.7 D
Power = 61 D

**(c)** See Section 2.5

## Chapter 3

**3.1** Fluid fills the Eustachian tube so that the pressure across the ear drum cannot be equalised.

**3.3** $I_{x} = I_{0}e^{-kx}$

**3.4** (a) 90 dB

(b) $10^{5}$

**3.5** (a) $10^{-12}$ W m$^{-2}$ from 1 kHz to 3 kHz

(b) 120 dB, 140 dB, no

(c) ~50 Hz to 14 kHz

**3.7** (a) $10^{16}$

(b) $10^{4}$ W m$^{-2}$

(c) from $10^{-10}$ to 100 W m$^{-2}$

**3.8** (a) For diffraction the aperture must be of the same size as the wavelength of the sound. It causes a change in direction of the sound;

(b) Speech frequencies are low;

(c) Consider the time period of a sound wave $(= \lambda f)$ compared to the time between separate sounds;

(d) Consider the path lengths for sound from each speaker to each ear.

**3.12** (a) See Section 3.1

(b) See Section 3.4

(c) $v = f \times \lambda$

$\lambda = v/f = 330/2600$

$\lambda/4 = 0.032$ m

length = 3.2 cm.

(d) intensity level = 10 log $I/I_{o}$

intensity level = 10 log $0.008/10^{-12}$

= 109 dB

**3.13** (a) See Section 3.2

(b) and (c) See Section 3.3

(d) Intensity level = 98 dB = 10 $\log_{10}$ $(I/10^{-12})$

$\therefore \dfrac{I}{10^{-12}}$ = antilog 9.8   $I = 6.31 \times 10^{-3}$

(e) Intensity of discomfort = $6.31 \times 10^{-2}$

$\therefore$ ten drills is the minimum number.

**3.14** (a) See Section 3.2 and Figure 3.4 (b)

(b) and (c) See Section 3.3d

(d) (i) $2.0 = 10 \log \dfrac{(I_{2})}{(I_{1})}$

$\dfrac{(I_{2})}{(I_{1})} = 1.58$

(ii) 10 steps = $10 \times 2$ dB = 20 dB

$\dfrac{(I_{2})}{(I_{1})} = 100$

$[1.58^{10} = 100]$

**3.15** See Section 3.4

# Chapter 4

**4.1** (a) You may have found that the range is too large and consequently the sensitivity is too low for this instrument, which is usually used to measure greater temperature differences.

(b) This will depend on how you did the experiment. Possible sources of error include the contact between the thermocouple and the body; the voltage measuring instrument; the reading of the meter scale; the process of calibration.

(c) A constant reference temperature must be maintained; good thermal contact with the body must be maintained.

(d) If you read off the temperature from tables, these will have been obtained by calibration against a reference thermometer of some kind. If you used a commercial thermometer to calibrate your thermocouple, this will have also been calibrated against a reference thermometer.

**4.3** There are illnesses which do not cause a rise in body temperature. Children, who are most at risk from rapid temperature rise, are not good at keeping the thermometer in contact with their bodies. The person's normal body temperature may differ from the accepted norm of 37 °C, so that changes may be hidden.

**4.4** (a) Taking drinks of water whilst running.

(d) 'He faces months of intensive rehabilitation'. His physical condition is still very weak and damaged (he will have to learn to walk with one leg). Also he has suffered brain damage, from which he can be expected to recover quite shortly.

**4.8** See Sections 4.2 and 4.3

**4.9** See Section 4.1

**4.10** See Thermography section.

# Chapter 5

**5.1** See Figure 5.17 below

| | Na$^-$ | Na$^+$ | K$^+$ | | Na | interstitial fluid |
|---|---|---|---|---|---|---|
| excess Na$^+$ | | | | and later | | |
| $-$ve potential | $\downarrow$ | $+$ve potential | $\uparrow$ | $-$ve potential | $\uparrow$ | axoplasm |
| | | | | | | cell membrane |
| 1. resting, polarised | | 2. action, depolarisation and reverse polarisation | | 3. recovery, repolarisation | | |

**5.2** (b) This is the pump for the systemic circulation, which includes organs and fine capillaries all over the body, so a strong contraction is required to provide the pressure.

**5.3** (a) It occurs at the same time as the much stronger ventricular depolarisation which hides it.

(b) About 1 per cent.

**5.5**　**(a)**　Immediate availability of information over an extended period, suitable for intensive care.

**5.6**　**(b)**　55, 120, 57–86, 67

**5.7**　**(a)**　Heavy plastic handles on the paddles, good insulation for the wiring and power unit, possibly insulating gloves and mat for the operator to stand on.

**(b)**　$20 \times 3000 \times 5 \times 10^{-3} = 300$ J

**(c)**　**(i)**　$150 \ \Omega$

**(ii)**　At the surface of the skin

**(d)**　see Figure 5.18 below

step-up transformer

**5.9**　**(a)**　pulse amplitude: 0.9 or 1.0 mV

**(b)**　$T = 0.7$ s

$$F = \frac{60}{0.7} = 86 \ \text{min}^{-1}$$

**(c)**　$T$ would get shorter
Flat part of trace would shrink

**(d)**　See Figure 5.19

**5.10**　See Section 5.1

---

## Chapter 6

**6.1**　Use the data and conversion factors in Table 6.1, e.g. arterial blood pressure, maximum value =

$$140 \times \frac{1.01 \times 10^5}{760} \ \text{Pa} = 18.6 \ \text{kPa}$$

**6.2**　**(a)**　$F = pA, \ \dfrac{l}{t} = \dfrac{v}{A}$

**(c)**　1.38 W

**6.3**　**(a)**　Operator practice will help to avoid some of the false results, such as those from the subject moving. Inaccuracies can originate from the fitting of the cuff; in the manual version, the reading of the manometer; in the electronic version, possibly the amplification of the signal.

**(b)**　Pressure in the vascular system varies with height. A drop of 10 cm below the position of the heart will increase the pressure by approx. 10 cm $H_2O$ or 7.5 mm Hg.

**6.4**　**(a)**　$10^{-2} \ \mu\text{V V}^{-1} \ \text{mm Hg}^{-1}$

**(b)**　0.60 mm

**6.5**　Piezoelectric, capacitance and resistance transducers can all measure displacement directly (explain how in your answer), inductance transducers measure velocity because the effect depends on movement, not simply position.

**6.7**　**(a)**　See Figure 6.11

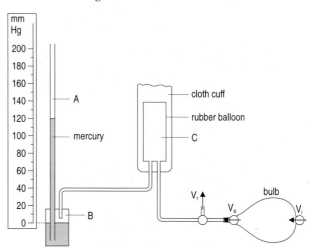

$V_r$ (release valve) allows air out
$V_e$ (exit valve) allows air out when bulb is squeezed but not in
$V_i$ (inlet valve) allows bulb to refill with air after squeezing.

**(b)**　A = 760 (mm Hg)
B = 880 (mm Hg)
C = 880 (mm Hg)

**(c)**　See Section 6.2

**(d)**　$p \ (= h\rho g) = 0.12 \times 1.36 \times 10^4 \times 10 = 1.6 \times 10^4 \ \text{Pa}$

**(e)**　extra pressure $(= h\rho g)$
$= 1.2 \ (\text{estimate}) \times 1.06 \times 10^3 \times 10 = 12\,700 \ \text{Pa}$
total pressure $= 16\,000 + 13\,000 = 29\,000 \ \text{Pa}$

**(f)**　blood pressure in head is lower as it is above the heart.
lowering the head increases blood pressure and hence improves blood flow to brain.

**6.8**　**(i)**　increased: $R_2, R_4$; decreased $R_1, R_3$

**(ii)**　$10 \ \Omega + 5\% = 10.5 \ \Omega$
$10 \ \Omega - 5\% = 9.5 \ \Omega$

**(iii)**　resistance of gauges very sensitive to temperature changes.
gauges can be connected in a (balanced bridge) network (in which all the resistances change by same amount for same temperature change) so that the effect of temperature change is

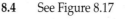

balanced out [or sensitivity increased by using four gauges].

## Chapter 7

**7.1**  **(a)** 0.616  **(b)** 38°

**7.2**  **(b)** 48 per cent

**7.5**  **(b)** The parallel collimated beam can be conveniently used in endoscopy. The monochromatic light is used for particular tissue interactions. Coherence of the phase of photons is also mentioned, this is important in holographic imaging, where interference patterns are produced but not in the treatments discussed here.

**(e)** Argon and helium-neon, the energy is supplied electrically. A high potential difference across the gas-filled tube causes the formation of ions by discharge.

**7.7**  **(a)** $10^5$ W mm$^{-2}$
**(b)** **(i)** $1.27 \times 10^{-3}$
**(ii)** 1.27 (over an area of $10^{-3}$ mm)
**(d)** **(i)** Turn your head or blink your eyes immediately.
**(ii)** Very little. The area of retina damaged would be too small to notice, and there would be no pain sensation.

**7.8**  **(a)** $\dfrac{\sin c}{l} = \dfrac{1.48}{1.60}$
critical angle, $c = 67.7°$
**(b)** $r = 90° - c = 22.3°$
$\dfrac{\sin \theta}{\sin r} = 1.60$
$\sin \theta = 1.60 \times \sin 22.3° = 0.61$ gives $\theta = 37.4°$

**7.9**  **(a)** See Section 7.1
**(b)** See Section 7.2

**7.10**  **(a)** and **(b)** See Section 7.3

**7.11**  **(a)** **(i)** $I = P/A$
$= 20/\pi \times (0.75 \times 10^{-3})^2$
intensity $= 1.1 \times 10^7$ W m$^{-2}$
**(ii)** intensity $= 1.1 \times 10^9$ W m$^{-2}$
**(b)** See Section 7.3

## Chapter 8

**8.1**  **(a)** **(i)** 2.72 mm, 0.97 mm, 2.53 mm **(ii)** 1.26 mm (half a wavelength)
**(b)** 10 cm (the pulse travels both ways)

**8.2**  **(a)** 0.26 – 0.44 or about 35 per cent.
**(b)** Almost 100 per cent reflection so there is no possibility of seeing tissue beyond.

**8.4**  See Figure 8.17

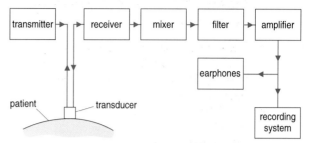

**8.5**  **(a)** The Doppler shift equations only show the difference, not which is larger.
**(b)** 22.5 mm s$^{-1}$

**8.6**  **(a)** 0.75 mm (half a wavelength).
**(b)** and **(c)** refer to Section 8.5

**8.7**  **(a)** (Specific) acoustic impedance
$Z_1 - Z_2$ large for empty bladder
Reflected signals strong
Insufficient transmitted
No tissue damage/can discriminate between soft tissues
**(b)** **(i)** 1 µs is $1 \times 10^{-6}$ s
3.5 MHz is $3.5 \times 10^6$ Hz
Moves in and out 3.5 mm (in 1 µs)
Graph: $3\frac{1}{2}$ cycles of sin/cos graph each about 0.3 µs on axis
**(ii)** $f \propto \dfrac{1}{\lambda}$ / $\lambda^{-1}$ /inversely proportional
High frequency means small wavelength which enables smaller detail to be resolved
**(iii)** Investigation: eye depth/length

**8.8**  See Section 8.1
See Section 8.3

**8.9**  See Sections 8.1 and 8.2

**8.10**  **(a)** See Section 8.4
**(b)** $\upsilon = \left(\dfrac{c\Delta f}{2f\cos\theta}\right) = \dfrac{1.5 \times 10^3 \times 800}{2 \times 4.0 \times 10^6 \cos 45°} = 0.21$ m s$^{-1}$

## Chapter 9

**9.1**  Refer to Figure 9.7

**9.2**  Complete Table 9.4

| Change | $E_{max}$ | $\lambda_{min}$ | shape of spectrum | characteristic lines | total intensity |
|---|---|---|---|---|---|
| increase $V_0$ | increases | | | | |
| increase $I_0$ | | unchanged | | | |
| increase $Z$ | | | | shifted to higher proton energy | |

205

**9.3** Use the conversion given to transform the values of photon energies given in Figures 9.5 (a) and 9.6, as is shown in Figure 9.4.

**9.4** (a) 4 kW
(b) $2.5 \times 10^{17}$
(c) $1.6 \times 10^{-14}$ J
(d) 0.0124 nm

**9.5** See Section 9.2

**9.6** See the preceding sub-section

**9.7** (a) 21 MW m$^{-2}$
(b) 0.21 MW m$^{-2}$
(c) 0.026 MW m$^{-2}$ (one eighth).

**9.9** Electrons have a charge so can be focused electrically or magnetically, X-rays are uncharged.

**9.11** Refer to Section 9.2
There is less simple scatter and photoelectric absorption of energy, with the higher energy photons, so there is greater penetration, greater deposition of energy at depth and smaller loss at or near the surface.
At MeV energy levels Compton effect predominates. This is independent of Z, so bone with high proton number, absorbs less of the energy than the mechanisms that predominate at lower energies.

**9.13** See Sections 9.1 and 9.2

**9.14** See Section 9.2 and Table 9.2

**9.15** See Section 9.2

**9.16** See Section 9.3

**9.17** See Sections 9.2 and 9.5

**9.18** See Section 9.5

## Chapter 10

**10.1** By taking potassium iodide tablets the thyroid of an irradiated person in which the radioactive iodine-131 collects can be 'flushed out' so that it contains only stable I-127.

**10.2** Activity from 96.8 per cent U-238 = 11.96 MBq kg$^{-1}$
Activity from 3.2 per cent U-235 = 2.53 MBq kg$^{-1}$
Total specific activity of enriched uranium = 14.49 MBq kg$^{-1}$

**10.3** The polonium also decays by α emission
$^{220}_{86}\text{Rn} \rightarrow {}^{216}_{84}\text{Po} + {}^{4}_{2}\text{He}$

**10.4** (a) $^{14}_{8}\text{O} \rightarrow {}^{14}_{7}\text{N} + \beta^{+} + \upsilon$
(b) $^{19}_{8}\text{O} \rightarrow {}^{19}_{9}\text{Fl} + \beta^{-} + \upsilon$

**10.5** (a) Magnetic and electric deflections: β great (high charge to mass ratio), α little (low charge to mass ratio), γ none (no charge).

(b) γ is em radiation ∴ $\upsilon = c$. Lower energy γ has a lower frequency $v$ and longer wavelength λ. $E = hv$, $c = v\lambda$

**10.6** (a) 1.8 m
(b) 21 mm of lead (four half-thicknesses).

**10.8** $^{A}_{Z}\text{X} + {}^{1}_{0}\text{n} \rightarrow {}^{A+1}_{Z}\text{X} + \gamma$

**10.9** $^{14}_{7}\text{N} + {}^{1}_{0}\text{n} \rightarrow {}^{14}_{6}\text{C} + {}^{1}_{1}\text{p}, \; {}^{6}_{3}\text{Li} + {}^{1}_{0}\text{n} \rightarrow {}^{3}_{1}\text{H} + \alpha({}^{4}_{2}\text{He})$

**10.10** Completions for table 10.3 neutron, α, neutron excess, neutron deficiency, β⁻, β⁺, no, yes.

**10.12** Refer to Table 10.6 and Section 10.4

**10.14** (a) See Section 10.1
(b) (i) 15.6 days
(ii) 4.7 days

**10.15** (a) $^{192}_{78}\text{Pt}$
(b) See Section 11.4
(c) $\lambda = \dfrac{\ln 2}{T\frac{1}{2}} = \dfrac{\ln 2}{75 \times 24 \times 3600} = 1.1 \times 10^{-7} \text{s}^{-1}$
[*alt*: $9.2 \times 10^{-3}$ day$^{-1}$]

(d) number of atoms in sample $= \dfrac{10 \times 10^{-3}}{192} \times 6 \times 10^{23}$
$= 3.1 \times 10^{19}$

(e) $\dfrac{\Delta N}{\Delta t}(=(-)\lambda N) = 1.1 \times 10^{-7} \times 3.1 \times 10^{19}$
$= 3.3 \times 10^{12}$ Bq [or s$^{-1}$]

(f) $N = N_0 e^{-\lambda t} = 3.1 \times 10^{19} \times e^{-1.06 \times 10^{-7} \times 30 \times 24 \times 3600} = 2.35 \times 10^{19}$

**10.16** See Section 10.4

**10.17** (a) See Section 10.1
(b) See section 10.5

## Chapter 11

**11.1** (a) Alpha's range in air is only a few centimetres, so it is unlikely to irradiate the body, as an external, sealed source. As a liquid or gas it can contaminate the body's surface, or be ingested and give a high local irradiation.

(b) The properties of X-ray and γ radiations are the same.

(c) The mass and charge of β have values between those of α and γ, which results in it having an ionising ability approximately midway between.

**11.2** This is a question for discussion, to see if you can identify all the possible factors.

**11.3** (a) 0.288 mSv (maximum dose permitted on any one occasion is 0.05 mSv)
(b) 0.719 mSv (maximum dose permitted in any one year is 0.5 mSv)

**11.4** 900 mJ

**11.6** (a) Approximately 64.5 μSv
(b) 64.5 hours
(c) about 3 days

**11.7** (a) The risk of radiation induced disease against the avoidance of toothache and other complications. Use the lowest possible doses, and have X-ray examinations only when really necessary.
(b) Their risk is set against the benefit of the wider population. In addition their actual exposure can be carefully monitored and controlled.
(c) These contain bone, muscles, etc., which are less radiosensitive.

**11.9** (a) See Section 11.2
(b) (i) quality factor 1:
$H(=QD) = 1 \times 20$ mSv = 20 mSv

Quality factor 20:
$H = 20 \times 20$ mSv = 400 mSv
Total dose equivalent = 20 + 400 = 420 mSv
(ii) $E = Dm$
energy = $40 \times 10^{-3} \times 60 = 2.4$ J

**11.10** (a) See Section 9.5
(b) (i) 2.4 m
(ii) 25.0 mm

## Chapter 12

**12.1** (a) Dark under the open window, light in all other parts
(b) Moderately dark over most of the film
(c) A dark patch, corresponding to the shape of the drop; the intensity will depend on which filter it landed on.

**12.2** (a) The RBE of neutrons is 20, so the dose equivalent is very high.
(b) α cannot penetrate the wrapping of the film, which keeps out the light. α emitters are not used in medical diagnosis or therapy.

**12.3** (b) Ionisation has taken place throughout the volume of the tube, so a finite time is required for the ions to travel to the electrons, before this pulse can be distinguished from that from the next burst of ionisation.

(c) 1 kBq.
(d) Because the counter is often taking measurements of the same order of magnitude as background radiation. The count may be corrected either by taking the background count in the absence of the source, and then subtracting it from the total count, or operating the counter within a shielded area.

**12.4** (b) (i) low energies, and small differences in energies can be detected
(ii) high count rates are possible (up to $10^4$ per second)
(iii) high density means a higher probability of the γ photon interacting with an atom in the scintillator

**12.6** See Section 12.4.

**12.7** (b) See Section 9.4.

**12.8** (a) Images can show thickening of the blood vessels, clots in them and other damage that can lead to fatal conditions.
(b) A small tube which can be inserted into the body, through a small incision.
(c) The first sentence says that the initial picture does not show blood vessels; in the last the doctor is using this image to help find the veins and arteries!

**12.9** See Section 12.1

**12.10** See Section 12.4

**12.11** See Sections 12.3 and 12.4

## Chapter 13

**13.1** Vary the number of turns per unit length, linearly.

**13.2** See Section 13.2

**13.3** See Section 7.2 for endoscopy and Sections 13.1 and 13.3

**13.4** (a) See Section 8.2. for ultrasound, Section 9.4 for X-rays
(b) See Section 13.3

# Appendix B: Technical information on the investigations

## Chapter 1

### The mechanics of jumping

This will require the participation of the P.E. or athletics department, if it is to be carried out with any degree of accuracy, though some rough estimates could be made with the use of laboratory measuring instruments, outdoors.

## Chapter 2

### Monocular and binocular vision

This is a series of simple investigations, requiring no equipment, which can be carried out at home.

## Chapter 3

### Sound and hearing

No specific investigations have been described in the text, but there are several possibilities, depending on the equipment available and the students' previous experience. Sound level meters are sold by most of the major school laboratory suppliers. An alternative way of measuring is to use a microphone with a CRO. Possible investigations include a person's range of hearing under different conditions (as suggested in question 3.6), noise levels in various situations and the absorption and reflection of sound by different materials.

## Chapter 4

### 1. Measuring body temperature with a thermocouple

It is intended that the apparatus required for this should be as simple as possible, in contrast with part (b) later in the chapter. A thermocouple can be made in the laboratory from 26 or 28 SWG copper and constantan wires. This will give a reasonable reading on a Scalamp galvanometer. An alternative is to use a thermistor in a bridge circuit with a high resistance (inexpensive bead thermistors are available from RS Components).

### 2. Temperature monitoring using sensors

This could be as 'high-tech' as available equipment and student imagination allows. Portability of equipment may be the main constraint.

## Chapter 5

### Measurement of ECG

Laboratory suppliers such as Philip Harris sell ECG interfaces which are easy to use with a computer and CRO, and permit simple investigations to be carried out.

## Chapter 6

### Measurement of blood pressure

This can be carried out with the traditional manual sphygmomanometer, or by one of the large range of electronic equivalents now marketed by laboratory suppliers.

## Chapter 7

### Optical fibres

No specific suggestions for investigations have been included in the text. There are a number of suppliers of non-coherent optical fibre, transmission kits and accessories, for example RS Components, Philip Harris and Irwin. These are useful to demonstrate transmission of signals. A hospital may be able to loan a fibre endoscope.

## Chapter 8

### The properties of ultrasound

A visit to a hospital will probably be neccessary to see ultrasonic imaging and Doppler measurements. Kits are available to investigate the nature of ultrasound (e.g. Unilab's Ultrasonics Kit), and to use it in measurement (e.g. Educational Electronics' Motion sensor which is a transmitter/receiver).

## Chapter 9

### 1. Half-value thickness

This investigation requires only a range of thicknesses of wood and lead and the standard cobalt-60 source in its storage box.

### 2. Shadows

This is a simple exercise which can be done at home with household items.

## Chapter 10

### 1. Half-life of radioisotopes

### 2. Range of radiations in air

These experiments could usefully be done at this stage, if students have not already performed them in their physics course. Details are not given in the chapter but may be found in standard physics texts.

## Chapter 11

### Background radiation

A range of everyday materials and substances can be measured using a standard laboratory G–M tube and a scalar or ratemeter. Suggestions include uranium, thorium and potassium salts, a gas mantle (these are coated with thorium oxide), granite, and brazil nuts!

## Chapter 12

### Film detection

This investigation makes use of the sealed laboratory sources, or uranium salts. It requires dental X-ray films which you may be able to beg from local dentists. The films and processing chemicals can be purchased from photographic suppliers; the processing can easily be carried out in a dimly lit laboratory.

Other possible practical work will depend on the availability of different detectors. It may be possible to obtain film badges and TLDs from a local hospital or laboratory for examination, though processing will not be easy to arrange. The performance of the Geiger counter at different voltages can be investigated, to determine the optimum conditions; details can be found in standard physics textbooks.

## Chapter 13

### Magnetic precession

This is simply modelled with a pair of bar magnets and an ordinary plotting compass.

# Index

Bold page references refer to an illustration, figure or table.